SCIENCE

A HISTORY IN 100 EXPERIMENTS

SCIENCE

A HISTORY IN 100 EXPERIMENTS

JOHN GRIBBIN
AND MARY GRIBBIN

WILLIAM
COLLINS

William Collins
An imprint of HarperCollins*Publishers*
1 London Bridge Street
London SE1 9GF

WilliamCollinsBooks.com

First published in Great Britain by William Collins in 2016

21 20 19 18 17 16
10 9 8 7 6 5 4 3 2 1

A catalogue record for this book
is available from the British Library

ISBN 978-0-00-814560-6

Edited and designed by Patricia Briggs

Printed in China by RRD Asia Printing Solutions.

Astronaut working on the
Hubble Space Telescope
(HST) during a routine
servicing mission.

CONTENTS

LIGO gravitational wave detector. Aerial photograph of the Livingston detector site for the Laser Interferometer Gravitational-Wave Observatory (LIGO). LIGO compares measurements between two detector sites 3000 kilometres apart, one near Hanford, Washington, USA, and the other near Livingston, Louisiana, USA. Each site is an L-shaped ultra-high vacuum system, four kilometres long on each side. Laser interferometers are used to look for small changes caused by gravitational waves. LIGO has been operating since 2002, with an advanced upgrade (aLIGO) operating since 2015. On 11 February 2016 it was announced that gravitational waves had been detected by LIGO. The signal was detected on 14 September 2015, and was the result of two black holes colliding.

INTRODUCTION

Science is nothing without experiments. As the Nobel Prize-winning physicist Richard Feynman said: 'In general, we look for a new law by the following process: First we guess it; then we compute the consequences of the guess to see what would be implied if this law that we guessed is right; then we compare the result of the computation to nature, with experiment or experience [observation of the world], compare it directly with observation, to see if it works. If it disagrees with experiment, it is wrong. In that simple statement is the key to science. It does not make any difference how beautiful your guess is, it does not make any difference how smart you are, who made the guess, or what his name is — *if it disagrees with experiment, it is wrong.*'[1]

Those words – *if it disagrees with experiment, it is wrong* – provide the simplest summary of what science is all about. People sometimes wonder why it took so long for science to get started. After all, the Ancient Greeks were just as clever as us, and some of them had both the curiosity and the leisure to philosophize about the nature of the world. But, by and large, with a few exceptions, that is all they did – philosophise. We do not intend to denigrate philosophy by this remark; it has its own place in the roll of human achievements. But it is not science. For example, these philosophers debated the question of whether a light object and a heavy object dropped at the same time would hit the ground at the same time, or whether the heavier object would fall more quickly. But they did not test their ideas by dropping objects with different weights from the top of a tall tower; that experiment would not be carried until the seventeenth century (although not, as we shall explain, by Galileo; see page 26). Indeed, it was just at the beginning of the seventeenth century that the English physician and scientist* William Gilbert (see page 24) first spelled out clearly the scientific method later summed up so succinctly by Feynman. In 1600, writing in his book *De Magnete*, Gilbert described his work, notably concerning magnetism, as 'a new kind of philosophizing', and went on: 'If any see fit not to agree with the opinions here expressed and not to accept certain of my paradoxes, still let them note the great multitude of experiments and discoveries … we have dug them up and demonstrated them with much pains and sleepless nights and great money expense. Enjoy them you, and if ye can, employ them for better purposes … Many things in our reasonings and our hypothesese will perhaps seem hard to accept, being at variance with the general opinion; but I have no doubt that hereafter they will win authoritativeness from the demonstrations themselves.'[2]

William Gilbert (1544–1603), English physician and physicist. In 1600 Gilbert published *De Magnete* (*Concerning Magnetism*), a pioneering study in magnetism, which contained the first description of the scientific method, and greatly influenced Galileo.

* The term 'scientist' was not coined until much later, but we shall use it for convenience to describe all the thinkers or 'natural philosophers' of centuries past.

In other words, *if it disagrees with experiment, it is wrong.* The reference to
'great money expense' also strikes a chord in the modern age, when scientific
advances seem to require the construction of expensive instruments, such as the
Large Hadron Collider at CERN, probing the structure of matter on the smallest
scale, or the orbiting automatic observatories that reveal the details of the Big
Bang in which the Universe was born. This highlights the other key to the

Robert Hooke's hand-crafted microscope.

relatively late development of science. It required (and requires) technology. There is, in fact, a synergy between science and technology, with each feeding off the other. Around the time that Gilbert was writing, lenses developed for spectacles were adapted to make telescopes, used to study, among other things, the heavens. This encouraged the development of better lenses, which benefited, among other things, people with poor eyesight.

A more dramatic example comes from the nineteenth century. Steam engines were initially developed largely by trial and error. The existence of steam engines inspired scientists to investigate what was going on inside them, often out of curiosity rather than any deliberate intention to design a better steam engine. But as the science of thermodynamics developed, inevitably this fed back into the design of more efficient engines. However, the most striking example of the importance of technology for the advancement of science is one that is far less obvious and surprises many people at first sight. It is the vacuum pump, in its many guises down the ages. Without efficient vacuum pumps, it would have been impossible to study the behaviour of 'cathode rays' in evacuated glass tubes in the nineteenth century, or to discover that these 'rays' are actually streams of particles – electrons – broken off from the supposedly unbreakable atom. And coming right up to date, the beam pipes in the Large Hadron Collider form the biggest vacuum system in the world, within which the vacuum is more perfect than the vacuum of 'empty' space. Without vacuum pumps, we would not know that the Higgs particle (see page 270) exists; in fact, we would not have known enough about the subatomic world to even speculate that such an entity might exist.

But we *know* that atoms and even subatomic particles exist, in a much more fundamental way than the Ancient Greek philosophers who speculated about such things, because we have been able (and, equally significantly, we have been willing) to carry out experiments to test our ideas. The 'guesses' that Feynman refers to are more properly referred to as hypotheses. Scientists look at the world around them, and make hypotheses (guesses) about what is going on. For example, they hypothesise that a heavy object and a lighter object dropped at the same time will hit the ground at different times. Then they drop objects from a high tower, and find that the hypothesis is wrong. There is an alternative hypothesis: that heavy and light objects fall at the same rate. Experiment proves that this is correct, so this hypothesis gets elevated to the status of a theory. A theory is a hypothesis that has been tested by experiment and passed those tests. Human nature being what it is, of course, it is not always so straightforward and clear cut. Adherents to the failed hypothesis may try desperately to find a way to shore it up and explain things without accepting the experimental evidence. But in the long run, the truth will out – if only because the die hards really do die.

Non-scientists sometimes get confused by this distinction between a hypothesis and a theory, not least because many scientists are guilty of sloppy use of the terminology. In everyday language, if I have a 'theory' about something (such as the reason why some people like Marmite and others don't) this is really just a guess, or a hypothesis; this is not what the word 'theory' means in science. Critics of Darwin's theory who do not understand science sometimes say that it is 'only a theory', with the implication 'my guess is as good as his'. But Darwin's theory of natural selection starts from the observed fact of evolution, and

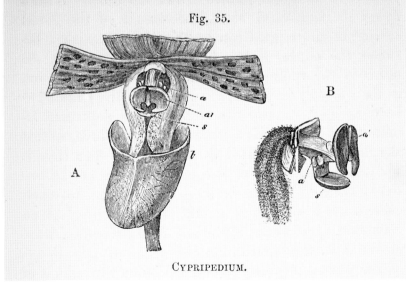

Charles Darwin's illustration, from his book *Fertilisation of Orchids*, of Cypripedium (slipper orchid, Paphiopedilum), beneath a photograph of an early variety of Sandford orchid cultivar.

explains how evolution occurs. In spite of what those critics might think, it is more than a hypothesis – not just a guess – because it has been tested by experiment, and has passed those tests. Darwin's theory of evolution by natural selection is 'only' a theory in the same way that Newton's theory of gravity is 'only' a theory. Newton started from the observed facts of the ways things fall or orbit around the Earth and the Sun, and developed an idea of how gravity works – gravity involving an inverse square law of attraction. Experiments (and further observations, which throughout this book we include under the heading 'experiments') confirmed this.

Gravity provides another example of how science works. Newton's theory passed every test at first, but as observations improved it turned out that the theory could not explain certain subtleties in the orbit of Mercury, the closest planet to the Sun, which orbits where gravity is strong – that is, where there is a strong gravitational field. In the twentieth century, Albert Einstein came up with an idea, which became known as the general theory of relativity, that explained everything that Newton's theory explained, but which also explained the orbit of Mercury and correctly predicted the way light gets bent as it passes near the Sun (see page 179). Einstein's theory is still the best theory of gravity we have, in the sense that it is the most complete. But that does not mean that Newton's theory has to be discarded. It still works perfectly within certain limits, such as in describing how things move under the influence of gravity in less extreme circumstances, in the so-called 'weak field approximation', and is fine for calculating the orbit of the Earth around the Sun, or for calculating the trajectory of a spaceprobe sent to rendezvous with a comet.

Contrary to what is sometimes taught, science does not proceed by revolutions, except on very rare occasions. It is incremental, building on what has gone before. Einstein's theory builds on, but does not replace, Newton's theory. The idea of atoms as little hard balls bouncing off one another works fine if you want to calculate the pressure of a gas inside a box, but has to be modified if you want to calculate how electrons jumping about within atoms produce the coloured lines of a spectrum of light. No experiment will ever prove the theories of Einstein or Darwin 'wrong' in the sense that they have to be thrown away or require us to start again, but they may be shown to be incomplete, in the way Newton's theory was shown to be incomplete. Better theories of gravity or evolution would need to explain all the things that the present theories explain, and more besides.

Don't just take our word for it. In his book *Quantum Theory*, Paul Dirac, possibly the greatest genius of the quantum pioneers, wrote: 'When one looks back over the development of physics, one sees that it can be pictured as a rather steady development with many small steps and superposed on that a number of big jumps. These big jumps usually consist in overcoming a prejudice … And then a physicist has to replace this prejudice by something more precise, and leading to some entirely new conception of nature.'[3]

All of this should be clear from the selection of experiments that we have chosen in order to mark the historical growth of science, starting with a couple of those rare pre-1600 exceptions that did amount to more than mere philosophising, and coming up to date with the discovery of what the Universe at large is made of. This choice is necessarily a personal one, and limited by the constraint of choosing exactly 100 experiments. There is so much more that we could have included. But one obvious feature of the story, which we realized as we were researching this book, is not a matter of personal choice, but another example of the way science works. Some of the experiments reported here come in clusters, with several in a similar area of science in a short span of time – for example, in the development of atomic/quantum physics. This is what happens when scientists succeed in 'overcoming a prejudice'. When a breakthrough is made, it leads to new ideas (new 'guesses', as Feynman would have said, but, crucially, *informed* guesses) and new experiments, which tumble out almost on top of each other until that seam is exhausted.

A problem for the non-specialist is that the information on which those guesses are based is itself based on the whole edifice of science, a series of experiments going back for centuries. The vacuum in the Large Hadron Collider has its origins in the work of Evangelista Torricelli in the seventeenth century (see page 30). But Torricelli could never have imagined the existence of the Higgs particle, let alone an experiment to detect it. The first steps in such a series are relatively easy to understand, even for non-scientists, not least thanks to the successes of science over the years. It is now 'obvious' to us that objects with different weight will fall at the same rate, just as it was 'obvious' to the ancients that they would not. But when it gets to the Higgs particle and the composition of the Universe, unless you have a degree (or two) in physics, it may be far from obvious that the story makes sense. At some level, things have to be taken on trust. But the key to that trust is that everything in the scientific world view is based on experiment, by which term we include observations of phenomena predicted by theories and hypotheses, such as the bending of light as it goes past the Sun (see page 179). If you find that some of the concepts described here fly in the face of common sense, remember what Gilbert said. They may 'seem hard to accept, being at variance with the general opinion'; but they 'win authoritativeness from the demonstrations [experiments] themselves'. And above all, *if it disagrees with experiment, it is wrong.*

THE UPWARD THRUST OF WATER

One of the first, and most famous, scientific experiments was carried out by Archimedes, who lived in the third century BC. Not much is known about Archimedes' personal life, but it seems that he was a relative of King Hieron II of Syracuse, in Sicily, and, after extensive travels, he settled down as the King's astronomer and mathematician. According to legend, King Hieron had a new crown, probably in the form of a laurel wreath, made for him from a bar of gold he supplied to the jeweller, to give as an offering to the gods in a temple. He suspected that the jeweller had kept some of the gold and mixed in cheaper silver instead to make up the same weight for the crown. This would be a doubly serious matter; not only would the king be cheated, but the gods might be offended at being given an inferior offering. So Hieron ordered Archimedes to find out if the crown was made of pure gold – without, of course, damaging it in any way. Archimedes had no idea how to do this, and worried about the problem for days. Then, when stepping in to a bath filled to the brim, he noticed how the water slopped over the side as it was displaced by his body. The story

An imaginative portrayal of the Greek mathematician and physicist Archimedes (287–212 BC) in his bath. Archimedes showed that an object immersed in a fluid is supported with a force equal to the displaced fluid's weight (Archimedes' principle).

has come down to us from Vitruvius, a Roman architect, in a book written two centuries after Archimedes had died. We do not know where he got it from, but this is where we get the image of Archimedes immediately realizing how to test the crown, and becoming so excited that he ran out into the street, naked and wet, shouting 'Eureka!' ('I have found it!').

What Archimedes had realized was that the volume of water displaced from the bath was equal to the volume of his body immersed in the water. As silver is less dense than gold, if the crown were made of a mixture of silver and gold it would have to be bigger than a crown made of pure gold in order to have the same weight. And he could measure the volume of the crown, without damaging it, by immersing it in water and seeing how much water was displaced.

Nobody knows exactly how Archimedes carried out the experiment. But the most likely method is based on an observation he described in his book, *On Floating Bodies*. There, Archimedes explained that the upward force (buoyancy) exerted on an object placed in water (or any other fluid) is equal to the weight of fluid that is displaced. This is now known as Archimedes' Principle. And, of course, the weight of water displaced will be proportional to the volume of water displaced.

The obvious way to use this to test the purity of the crown, as Archimedes must have realized, would be to balance the crown against exactly the same weight of pure gold on a beam balance above a tank of water. Then, the balance is lowered until the crown and the pure gold sample are immersed in the water, while the balance arm stays above it. If both objects are made of pure gold, they will each displace the same volume (and therefore the same weight) of water, experience the same buoyancy force, and stay in balance. But if the crown is less dense than gold it will have a bigger volume, displace more water, and be more buoyant than the pure gold, so the balance will tip down on the side of the gold. The beauty of this experiment is that you don't actually have to measure the volume of the crown, or the volume of water that it displaces; you just watch to see if the balance tilts.

That, it seems, is exactly what happened. Archimedes did the experiment (or something very similar) and found that the jeweller had indeed cheated the king. About five centuries after Vitruvius, the story was re-told in a Latin poem 'Carmen de ponderibus et mensuris' which described the use of such a hydrostatic balance, and in the twelfth century a manuscript called 'Mappae clavicula' gave detailed instructions on how to make weighings in this way to calculate the proportion of silver in the adulterated crown.

Archimedes' Principle also explains why a ship made of steel can float. A solid lump of steel displaces a relatively small amount of water, much less than its own weight, and sinks. But if the same amount of steel is spread out in the shape of a boat, or even a simple bowl (like a coracle), a larger volume of water is displaced, weighing more than the weight of the steel, resulting in a large enough upwards force to make the boat float.

MEASURING THE DIAMETER OF THE EARTH

The first scientific attempt to measure the size of the Earth was made by a Greek polymath, Eratosthenes of Cyrene (276–194 BC), who was in charge of the Library of Alexandria in the third century BC. He was a contemporary and friend of Archimedes. His experiment involved some observations of his own, made in Alexandria, but combined with evidence from a far away place, the city then known as Syene (now Aswan), which he had never visited.

Eratosthenes learned that each year on the day of the summer solstice, when the Sun is at its highest in the sky, it was exactly overhead at Syene, south of Alexandria. Travellers told how the reflection of the Sun could be seen at the bottom of a deep well in Syene on that day. Even at the summer solstice, the Sun is not directly overhead at Alexandria, because, as Eratosthenes appreciated, the Earth is round. So he made careful measurements of the difference between the angle made by the Sun and the vertical at the time of the solstice, working out that this corresponded to one-fiftieth of a circle, or 7° 12′ of arc. Simple geometry told him that this meant that the distance from Alexandria to Syene was one-fiftieth of the circumference of the Earth, assuming (which is not quite true) that Syene lies due south of Alexandria.

The distance from Syene to Alexandria was well known even in Eratosthenes' day (it is about 800 kilometres in modern units). Egyptian records gave the distance as 5,000 stades, and Eratosthenes checked this by asking camel train drivers how long it took them to make the journey (some sources say he hired a man to pace out the distance; but this may be apocryphal). This gave him a figure of 694 stades per degree, which he rounded off to 700. Multiplying by 360 gave him the circumference of the Earth – 252,000 stades (he could have just multiplied 5,000 by 50 to get the 'answer' 250,000, but apparently he did it the hard way).

So what is this in modern units? Unfortunately for us, the Greeks and Egyptians used slightly different stades, but the likelihood is that Eratosthenes, being Greek, used the Greek measurement, where one stade corresponds to 185 metres, which gives a circumference of 46,620 kilometres, only 16.3 per cent too big. In the unlikely event that he used the Egyptian measurement, with one stade corresponding to 157.5 metres, he would have come up with a figure of 39,690 kilometres, just a bit too small (less than 2 per cent smaller than the actual distance, 40,008 kilometres). Either way, it is impressive.

That was by no means the only impressive achievement of Eratosthenes. He used the information he found in the books in the Library of Alexandria to produce a three-volume book of his own in which he mapped and described the entire known world. He used grids of overlapping lines, like modern lines of latitude and lines of longitude, to locate places, and invented many of the terms still used by geographers today. More than four hundred cities were named and

Eratosthenes (*c.* 276–194 BC).

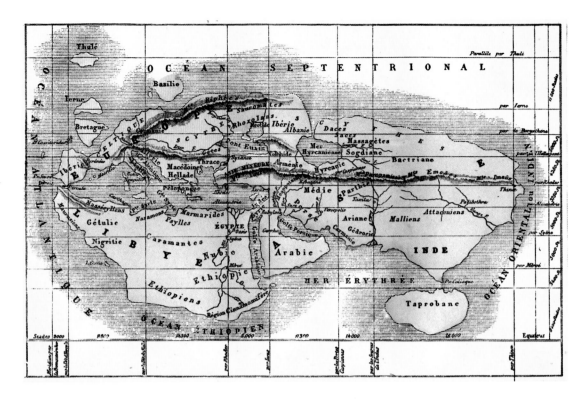

The World by Eratosthenes. 1886 replica of a map of the known world according to the Ancient Greek geographer, mathematician and astronomer Eratosthenes.

located in the book. Unfortunately the book itself, called *Geographika*, was lost, but parts of it have been reconstructed from references to it in other works. Book Two of *Geographika* included Eratosthenes' estimate of the size of the Earth. According to Ptolemy, Eratosthenes measured the tilt of the Earth's axis, which is related to the measurement of the circumference, very accurately, getting a value of $^{11}/_{83}$ of 180°, which is 23° 51' 15". He also worked out a calendar that included leap years, and he tried to establish a chronology of literary and political events going back to the siege of Troy.

Eratosthenes was very much an all-rounder, so much so that he had the nickname 'Beta', because he was second best at everything, according to his contemporaries. The Greek geographer Strabo, who lived from about 64 BC to AD 24, described Eratosthenes as the best mathematician among the geographers, and the best geographer among the mathematicians. In mathematics, he is known for a technique called 'the sieve of Eratosthenes', used to find prime numbers. This simple method, which he invented, involves making a list (or grid) of all the numbers up to the biggest one you are interested in (for example, 1 to 1000). Then, you cross off from the list all the multiples of 2, the first prime number (4, 6, 8 and so on, but not 2 itself), and check that the next lowest number not crossed off is prime (if it isn't, you have made a mistake!). If it is, cross off all the multiples of that number (but not the number itself), and so on. Once you get to the end of the list, the numbers that have not been crossed off form the list of primes.

THE EYE AS A PINHOLE CAMERA

After the decline of classical civilization and before the European Renaissance, scientific knowledge was preserved and improved in the Arabic world. Greek texts were translated into Arabic and later from Arabic into Latin, which is how they became known to Europeans. But the Arabs also carried out original scientific work. The greatest scientist of the Middle Ages, the 'Arabic Newton', was Abu Ali al-Hassan ibn al-Haytham, known for short as Alhazen, who lived from about 965 to 1040 and carried out experiments in optics on either side of the year 1000. His influential book was published in Europe in Latin as *Opticae Thesaurus* (*The Treasury of Optics*) in 1572, five centuries after his death. It was a major influence on the 'natural philosophers' who started the scientific revolution in Europe.

Alhazen's key insight was that sight is not the result of some influence reaching out from the eyes and sensing the world outside, but is caused by light entering the eye from outside. In his own words, 'from each point of every coloured body, illuminated by any light, issue light and colour along every straight line that can be drawn from that point'. This was not an entirely original idea. Philosophers had discussed whether vision was caused by an outward influence (emission) or an inward influence (intromission) since the time of Euclid and Aristotle. But Alhazen put together a complete, coherent package of ideas which he then proved correct by experiments based on the idea of a 'camera obscura' (literally, a 'dark room'; the Latin term is the source of our modern word camera). In a dark room with a heavily curtained window, if a tiny hole is made in the curtain on a sunny day an image of the outside world will be projected, upside down, on the wall opposite the window. The phenomenon had been known to the ancients, but Alhazen was the first person to describe it clearly and explain what is going on.

Alhazen realized that light travels in straight lines. Light from the top of a tree in the garden outside the window of the camera obscura will go through the hole in the curtain to the bottom of the wall opposite. Light from the base of the tree will go through the hole and up to the top of the wall. Straight lines from other points on the tree, and

Alhazen's representation of the eye as a 'camera'.

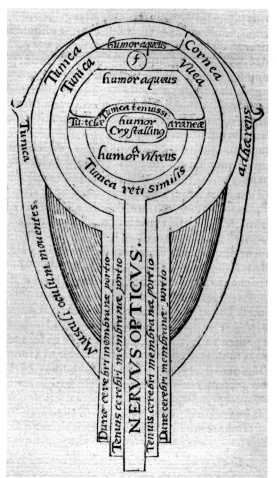

Abu Ali al-Hassan ibn al-Haytham (known as Alhacen, or Alhazen) (965–1040).

from other objects outside the window, go through the hole in straight lines to corresponding places on the wall to make the image.

He might have stopped there. Before Alhazen, those philosophers who thought about such things at all, such as Euclid and Aristotle, usually stopped at this stage, without actually doing experiments to test their ideas. They tried to persuade people that they were right by logic and reason, without getting their hands dirty (Archimedes, of course, was a notable, but rare, exception). What made Alhazen a real scientist was that he went a stage further. It was one thing to show how a camera obscura worked, but something else to prove that the eye works in the same way. A thousand years ago, many people would have assumed that living things were not subject to the same rules as inanimate objects. To test whether this was so, he took an eyeball from a bull, and carefully scraped away at the back of it, thinning it down until he could see on the back of the eyeball an image of what was in front of the eye, almost exactly like a tiny camera obscura. He had proved that light travels in straight lines, shown how a camera obscura works, and established that no mysterious life force is needed to explain vision, just the same physical laws that apply to non-living things. And he had done so using what became known (eventually) as the scientific method – thinking up ideas (hypotheses) about how the world works based on observation, then testing those ideas by experiment. Today, an idea that passes the experimental test is upgraded to the status of a theory, while those that fail the experimental test are discarded. As the twentieth-century physicist Richard Feynman pithily put it, 'if it disagrees with experiment then it is wrong'. Because he understood this and put it in to practice, Alhazen was arguably the first modern scientist.

Alhazen did much more than this. He wrote on a variety of scientific and mathematical topics, and his optical work alone filled seven books. He realized that light does not travel at infinite speed, although it is very fast, and he explained the illusion that a straight stick looks bent when one end is placed in water because light travels at different speeds in water and in air. He studied lenses and curved mirrors, working out how the curvature makes them focus light. But his place in history has been secured by how he worked as much as by what he studied. It was the true beginning of experimental science.

Nᵒ. 4 DISSECTING THE HUMAN BODY

The scientific Renaissance began in the middle part of the sixteenth century, and a significant marker is the year 1543, when Copernicus published his famous book *De Revolutionibus Orbium Coelestium* (*On the Revolution of Celestial Bodies*), displacing the Earth from its supposed special position in the Universe, and Andreas Vesalius published *De Humani Corporis Fabrica* (*On the Structure of the Human Body*), going some way towards displacing humankind from a supposed special position in the animal world. Copernicus's story is well known, and he did not, strictly speaking, carry out experiments. But Vesalius is less well known, and deserves more attention than he often gets. He did carry out experiments – on human bodies.

Vesalius was born in Brussels in 1514, but carried out his important work at the University of Padua (where he was Professor of Anatomy) in the late 1530s and early 1540s. Before that time, when human dissections were carried out (which was not very often), the actual cutting was done by barber-surgeons, who were little more (arguably less) than butchers. The professor would stand at a safe distance (literally without getting dirty) and lecture to students about what was being uncovered, using imagination as well as actual evidence. Vesalius changed all that. He carried out the dissections himself, showing as well as telling the students what was going on, and developing a much better understanding of the human body. He was helped by the civil authorities in Padua – in particular, the judge Marcantonio Contarini, who not only supplied him with the bodies of executed criminals but would time the executions to fit in with Vesalius's need for a fresh cadaver for a lecture. This was in marked contrast to his time as a student in Paris, where Vesalius (like his fellow medical students) had been reduced to grave-robbing to get specimens for his studies.

Before Vesalius, the accepted understanding of human anatomy had been handed down since ancient times, and was based on the work of the Romano-

Andreas Vesalius (1514–1564).

Greek physician Claudius Galenus (known as Galen). In the Middle Ages in Europe, it was thought that the ancients had been much wiser than contemporary people, and that they had superior knowledge which could not be emulated, much less exceeded. But this was wrong. Galen was an enthusiastic dissector, but most of his work was carried out on dogs, pigs, and monkeys, because human dissection was infra dig in the second century AD. So his description of the human body was often wildly inaccurate.

The big contribution Vesalius made was not just to improve the understanding of human anatomy, but to stress the importance of using the evidence in front of you and your own experiments to find things out, instead of relying on the supposedly superior wisdom of the ancients. This, of course, echoed what was happening in astronomy at the same time. Vesalius, who once wrote, 'I am not accustomed to saying anything with certainty after only one or two observations', used to carry out

'parallel dissections' in which an animal body and a human body are dissected side by side, to highlight the anatomical differences between them, explicitly correcting Galen's errors.

Vesalius also used highly skilled artists to prepare large diagrams to use in his lectures (a sixteenth-century equivalent of PowerPoint), and six of these were published as *Tabulae Anatomica Sex* (*Six Anatomical Pictures*) in 1538. He drew three of the illustrations himself, but the other three were made by John Stephen of Kalkar (Jan Stephen van Calcar), a pupil of Titian. Stephen is also thought to have been the main illustrator for Vesalius's masterwork, the *Fabrica*, which appeared in seven volumes in 1643. But Vesalius' pioneering activity did not stop there. The *Fabrica* was a book for experts – other professors and doctors. In order to make his work accessible to students, and even to educated laymen, he produced alongside it, and published in the same year, a summary officially titled *De Humani Corporis Fabrica Librorum Epitome* (*Abridgement of the Structure of the Human Body*) but known as the *Epitome*. And all before he was 30. He then gave up teaching, and spent the rest of his career practising medicine as a Court Physician, first to the Holy Roman Emperor Charles V and then to Charles's son, Philip II of Spain (who later sent the Spanish Armada against England).

This change of career may have been caused by the opposition to his ideas from some of his peers, even in Padua. Jacobus Sylvius, a physician of the old school based in Paris, said that Vesalius was mad and that any advance of anatomical knowledge beyond Galen was impossible. It was more likely, he said, not that Galen was wrong, but that the human body had changed since his time. In 1543, science still had a long way to go.

N°. 5   MEASURING THE MAGNETIC FIELD OF THE EARTH

If you were looking for a key date to mark the transition from a superstitious and mystical view of the world to the scientific study of our surroundings, you could do worse than pick 1600, the year in which the first book based solely on scientific experimentation was published. The book was called *De Magnete Magneticisque Corporibus, et de Magno Magnete Tellure* (*Concerning Magnetism, Magnetic Bodies, and the Great Magnet Earth*), usually shortened to *De Magnete*, and it was the work of an Elizabethan physician/scientist who had spent years studying magnetic phenomena.

William Gilbert had been born in 1544, studied at Cambridge and eventually became a Court Physician, first to Elizabeth I then to James I of England and VI of Scotland. As a wealthy gentleman, he was able to indulge his passion for science as an amateur, but reportedly spent £5,000 of his own money on this 'hobby'. He died, probably of bubonic plague, in 1603.

Some of the most important experiments carried out by Gilbert concerned the magnetism of the Earth. At the time, seafarers were opening up the exploration of the world, and the magnetic compass was an invaluable tool, although nobody understood how it worked. Gilbert discussed the behaviour of compass needles with ships' captains and navigators, and disproved by

experiment many superstitions, such as the idea that a magnetic compass could be desensitized by rubbing it with garlic, or even by garlic breath. He then worked with naturally occurring magnetic rocks called lodestones, shaped into magnetized spheres that he called *terrellae* (meaning little Earths). He studied the magnetism of these spheres with magnetized needles which could be moved around the spheres. Gilbert showed that these behaved like compass needles at different places on Earth, and concluded that the Earth had an iron core which behaved like a bar magnet, with a North Pole and a South Pole. Before he carried out these experiments, philosophers had argued that compass needles pointed north because they were attracted to the Pole Star, or, alternatively, that there was a large magnetic island at the north geographical pole.

All of this represented a dramatic scientific leap forward, which is so obvious to modern eyes that it is hard to appreciate its revolutionary nature at the time. Gilbert regarded his terrellae as models of the real Earth, and accepted that the results he obtained – for example, in the way the angle of dip of a magnetized needle depends on its place on the magnetized sphere – could be scaled up to tell us what the Earth itself is like. He was extrapolating from models to the world at large, a key feature of science in subsequent centuries.

Title page of the second edition of William Gilbert's *De Magnete*, published in 1628.

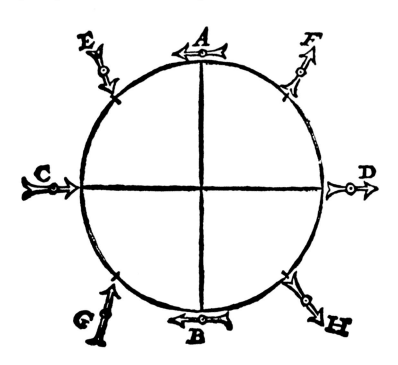

William Gilbert's illustration of the angle of dip of a magnetic field surrounding the Earth; the line AB is the equator, C is the North Pole, and D is the South Pole.

As a result of these experiments, Gilbert was the first person to appreciate that, because magnetic opposites attract, the end of a magnet that points northwards (towards the north magnetic pole of the Earth) ought to be called the south pole. (In modern language, scientists sometimes refer to the 'north-seeking' pole and the 'south-seeking' pole of a magnet to avoid this confusion.) Gilbert said: 'All who hitherto have written about the poles of the loadstone, all instrument-makers, and navigators, are egregiously mistaken in taking for the north pole of the loadstone the part of the stone that inclines to the north, and for the south pole the part that looks to the south: this we will hereafter prove to be an error. So ill-cultivated is the whole philosophy of the magnet still, even as regards its elementary principles.'

Indeed, it was Gilbert who introduced terms such as 'magnetic pole' and 'electric force' into the language. He was the first person to realize that magnetism and electricity (a word he invented) are separate phenomena, and his work on magnetism was not improved upon for two centuries, until the work of Michael Faraday.

Gilbert's book caused a sensation in its day, and was highly influential. Galileo was one of its readers, and commented favourably on it in a letter to a friend. Indeed, Galileo described Gilbert as the founder of the scientific method. In Gilbert's own words: 'In the discovery of secret things, and in the investigation of hidden causes, stronger reasons are obtained from sure experiments and demonstrated arguments than from probable conjectures and the opinions of philosophical speculators.' That is science in a nutshell, and in his book Gilbert was careful to spell out every detail of his experiments, so that other people could carry them out and see the results for themselves. But he cautions whoever does this 'to handle the bodies carefully, skilfully and deftly, not heedlessly and bunglingly; when an experiment fails, let him not in his ignorance condemn our discoveries, for there is naught in these Books that has not been investigated and again and again done and repeated under our eyes'.

Nº. 6 MEASURING INERTIA

G alileo Galilei is famous for an experiment he did not carry out – but it was a real experiment, inspired by his work. He was Professor of Mathematics in Padua from 1592 until 1610, and during that time he worked in mechanics and astronomy as well as mathematics.

To put Galileo's achievements in perspective, at the time he was working there were many people – educated people – who thought that a bullet fired horizontally from a gun, or a ball fired from a cannon, would fly a certain distance in a straight line, then stop and drop vertically to the ground. It was Galileo who first appreciated that the trajectory followed when a bullet is fired

Nineteenth century illustration showing a glamourised version of Galileo's experiment rolling balls down inclined planes. Although the scene depicted is fictional, Galileo really did carry out such experiments.

from a gun, or when an object such as a ball is thrown up in the air, is a parabola – and he carried out tests to prove this.

Among the many experiments he carried out in the years around the turn of the century there was a series of studies in which he rolled balls with different weights down inclined planes. He timed how quickly the balls moved using his pulse, and reached two important conclusions. The first was that the 'natural' state of a ball rolling off the slope was to continue horizontally (literally 'towards the horizon'), unless it was stopped by friction. Without friction, he reasoned, the ball would roll on forever. This was an early insight into what Isaac Newton, following Robert Hooke, developed as his 'First Law' of mechanics, that an object stays at rest or moves in a straight line at a steady speed unless it is acted upon by an outside force.

Galileo's second discovery was that the speed with which the balls rolled down the slope did not depend on their weight. For any particular slope, it took the same amount of time for any of the balls to get from the top to the bottom. This applied no matter how steep he made the slope. So he concluded – without actually dropping things vertically – that, apart from the effects of wind resistance, all falling objects would accelerate downwards at the same rate.

This infuriated some of his colleagues, philosophers of the old school who believed that Aristotle, who said that heavy objects fall faster than light objects, could not be wrong. So, in 1612, two years after Galileo moved from Padua to Pisa, one of them really did drop two weights from the leaning tower in a public demonstration intended to prove that Aristotle was right. The balls hit the ground very nearly at the same time, but not exactly. The Aristotelians said that this proved Galileo was wrong. But Galileo had an answer: 'Aristotle says that a hundred-pound ball falling from a height of one hundred cubits hits the ground before a one-pound ball has fallen one cubit. I say they arrive at the same time. You find, on making the test, that the larger ball beats the smaller one by two inches. Now, behind those two inches you want to hide Aristotle's ninety-nine cubits and, speaking only of my tiny error, remain silent about his enormous mistake.'[4]

Among other things, this true story highlights the power of the experimental method. If you carry out an experiment honestly, it will tell you the truth, regardless of what you want it to tell you. The Aristotelians wanted to prove Galileo was wrong, but the experiment proved he was right – within, as we would now say, the limits of possible experimental error.

By 1612 Galileo was nearly 50, and his days as an experimental physicist were essentially over. His famous clash with the Church authorities in Rome did not take place until the 1630s, and led to his spending his final years, from 1634, under house arrest at his own home (a relatively lenient sentence considering that he had been forced to confess to heresy). There, he summed up his life's work on mechanics and promoted the scientific method pioneered by Gilbert in a great book, *Discourses and Mathematical Demonstrations Concerning Two New Sciences*, usually known as *Two New Sciences*, published in 1638 in Holland. The book was enormously influential, the first real scientific textbook, and an inspiration to scientists across Europe – except, of course, in Catholic Italy, where it was banned. As a direct result, from being a leading light in the scientific renaissance, Italy became a backwater, while the real progress was made elsewhere.

Nº. 7 ── CIRCULATION OF THE BLOOD

ven after the publication of the *Fabrica* (see page 21), in the second half of the sixteenth century and early in the seventeenth century there was still strong opposition to the idea that classical teachers such as Galen could be wrong. So, although the English physician William Harvey was born in 1578 and studied the circulation of the blood in the early decades of the seventeenth century, he did not publish his discoveries until 1628, by which time he had gathered an overwhelming weight of evidence to support his ideas. (He did, however, give lectures on his work in 1616.) The result was a book, *De Motu Cordus et Sanguinis in Animalibus* (*On the Motion of the Heart and Blood in Animals*, known as *De Motu*), which presented an open-and-shut case based on a series of genuinely scientific experiments carried out over the previous two decades. All this was done in Harvey's spare time as a successful physician who had studied in Cambridge and Padua and, like William Gilbert, became (in 1618) one of the Court Physicians to James I, and later personal physician (a much more important post) to Charles I. Both Williams were contemporaries of another William, Shakespeare, who died in 1616.

Before Gilbert, following the teaching of Galen, it was thought that veins and arteries carried two different kinds of blood. One kind, supposedly manufactured in the liver, was thought to be carried through the veins to nourish the tissues of the body, getting used up in the process and being replaced by new blood from the liver. The other kind of blood was thought to be carried in the arteries, conveying a mysterious 'vital spirit' from the lungs to the tissues of the body.

As with Gilbert (see page 24), the way in which Harvey worked and presented his results was as important as the discoveries themselves. He did not base his ideas on abstract philosophising, but on direct measurements and observations. The key insight came when he measured the capacity of the heart and, by taking a typical pulse rate, worked out how much blood it was pumping each minute. He found that, in modern units, a human heart pumps about 60 cubic centimetres with each beat, adding up to nearly 260 litres in an hour. That much blood would weigh three times as much as a human, so it was clearly impossible that it was all being manufactured in the liver (or anywhere else) every hour. The only alternative was that there was a lot less blood, and that it was continuously circulating around the body, out from the heart through the arteries and back through the veins. An equivalent system circulates blood between the lungs and the heart, carrying not 'vital spirit' but oxygen. All this was born out by Harvey's observations of the tiny valves in veins (discovered by one of Harvey's teachers in Padua, Hieronymous Fabricius) which allow venous blood to flow towards the heart, but not away from it.

Having reached this conclusion by observation, Harvey then established his case with a series of experiments, one of which stands out for its simplicity and clarity. If he was right, there must be a connection between arteries and veins. As arteries lie deeper below the surface of the skin than veins, he tested this by tying a cord (ligature) around his own arm, tight enough for it to cut off the flow of blood in his veins, but not in his arteries. As blood continued it flow

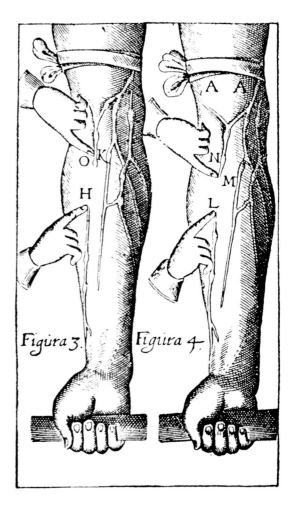

Woodcut from William Harvey's book, *De Motu Cordis et Sanguinis in Animalibus*. The illustrations show the valves in the superficial veins of the forearm. On the left, a finger has been passed along the vein from O to H (away from the heart). The stretch of vein is emptied and remains so because of the valve at O.

Title page from William Harvey's _De Motu_.

from the arteries into the blocked-off veins, the veins behind the ligature swelled up dramatically. He also pointed out that arteries near the heart are thicker than those further away from it, because they have to be strong in order to cope with the powerful pumping action of the heart.

There were still elements of mysticism in Harvey's thinking, and he saw the heart as not merely a pump but a place where the blood was made perfect by 'the foundation of life, and author of all'. It was René Descartes who took the next step, drawing on Harvey's work, and said, in 1637, that the heart is simply a mechanical pump.

Although his book caused great interest in England, Harvey's ideas about the circulation of the blood were not fully accepted during his lifetime, except by pioneers such as Descartes. One reason was that blood-letting was then (and would long remain) a treatment for illness, and the rationale for such treatment would be undermined if there was a limited amount of blood in the body.

Harvey died in 1657 and soon afterwards the development of the microscope (see page 35) made it possible to see the tiny connections between veins and arteries, establishing once and for all that Harvey had been correct.

WEIGHING THE ATMOSPHERE

In the early 1640s, the Italian Evangelista Torricelli investigated the problem that water could not be pumped up (by a suction pump) from a well more than about 30 feet (roughly 9 metres) deep. The way these pumps work is similar to the way it is possible to suck water into a bicycle pump if the open end is placed below the surface of the water. If you had a very long bicycle pump standing upright in a swimming pool, you would be able to suck water up to just over 30 feet, but no further, no matter how hard you pulled on the handle. Torricelli reasoned that the weight of the air pressing down on the surface of the water in the well could push the air in a pipe up this far, but no further. So he set out to test the idea using a denser liquid, mercury, instead of water. Mercury is roughly fourteen times denser than water, so Torricelli worked out that a column of water 30 feet high must exert the same pressure at its base as a column of mercury a bit more than 2 feet (more than 60 centimetres) high. He found that if a glass tube sealed at one end and full of mercury was stood upright with its open end in a dish of mercury, the level of the mercury in the tube would fall to 30 inches (76 centimetres), leaving a gap above the top of the mercury in the tube, and matching his calculation. The gap contained nothing at all, and became known as the Torricelli Vacuum.

Torricelli noticed that the exact height of the column of mercury in his tubes changed from day to day, and he realized that this was because the

pressure of the atmosphere weighing down on the mercury in the dish was
changing. He had invented the barometer. Torricelli died in 1647, but his
discoveries were taken up and developed by the Frenchman Blaise Pascal, who
studied the way the pressure of the air, measured by this kind of early
barometer, varied with the weather. Another Frenchman, René Descartes,
visited Pascal in 1647, and suggested that it would be interesting to take a
barometer up a mountain, to find out how the pressure of the air changed with
altitude. Pascal lived in Paris, but his brother-in-law Florin Périer lived near a
mountain, the Puy-de-Dôme, and in 1648 Pascal persuaded him to do the
experiment. Périer write to Pascal to describe what happened:

'The weather was chancy last Saturday … [but] around five o'clock that
morning … the Puy-de-Dôme was visible … I decided to give it a try. Several
important people of the city of Clermont had asked me to let them know when I
would make the ascent … I was delighted to have them with me in this great work.

… at eight o'clock we met in the gardens of the Minim Fathers [monastery],
which has the lowest elevation in town … First I poured 16 pounds of quicksilver

Imaginative illustration of
Florin Périer measuring the
air pressure as he ascends
the Puy-de-Dôme, a volcanic
mountain in France with a
height of 1464 metres.

... into a vessel ... then took several glass tubes ... each four feet long and hermetically sealed at one end and opened at the other ... then placed them in the vessel ... the quick silver stood at 26" and 3½ lines above the quicksilver in the vessel ... I repeated the experiment two more times while standing in the same spot ... [it] produced the same result each time ...

I attached one of the tubes to the vessel and marked the height of the quicksilver and ... asked Father Chastin, one of the Minim Brothers ... to watch if any changes should occur through the day ... Taking the other tube and a portion of the quick silver ... I walked to the top of Puy-de-Dôme, about 500 fathoms higher than the monastery, where upon experiment ... found that the quicksilver reached a height of only 23" and 2 lines ... I repeated the experiment five times with care ... each at different points on the summit ... found the same height of quicksilver ... in each case ...'[5]

Equally importantly, the priest at the bottom of the mountain reported that the reading on his barometer had not changed during the day. There was less weight of air pressing down at the top of the mountain than at the bottom. So the experiment revealed that the atmosphere gets thinner as you go higher, and suggests that if you go high enough it will thin out entirely, with a vacuum above it, like the vacuum above the mercury in Torricelli's tubes. Pascal then carried out a mini-version of the experiment by carrying a barometer up about 50 metres to the top of the bell tower at the church of Saint-Jacques-de-la-Boucherie. The mercury dropped by two 'lines'. Many people, including Descartes, refused to accept Pascal's interpretation of the evidence, and insisted that there must be some invisible substance filling the 'empty' space in the tubes and (presumably) the space above the atmosphere. But further experiments eventually proved that Pascal was right (see page 186).

Nᴼ· 9   RESISTING THE SQUEEZE

Following the experiments of Torricelli, Pascal, and his brother-in-law, the study of the vacuum became one of the hottest topics in science. In order to investigate this phenomenon, scientists needed very efficient pumps that could suck air out of glass bottles and other vessels. These pumps were hi-tech by the standard of the day – the seventeenth-century equivalent of modern particle accelerators such as the Large Hadron Collider. The very best air pumps available in the 1660s were made by the British scientist Robert Hooke, who was working at the time as an assistant to Robert Boyle. Boyle was a pioneering scientist (he helped to found the Royal Society) inspired by the work of Galileo, and once said that in investigating the world 'we assent to experience, even when its information seems contrary to reason'.

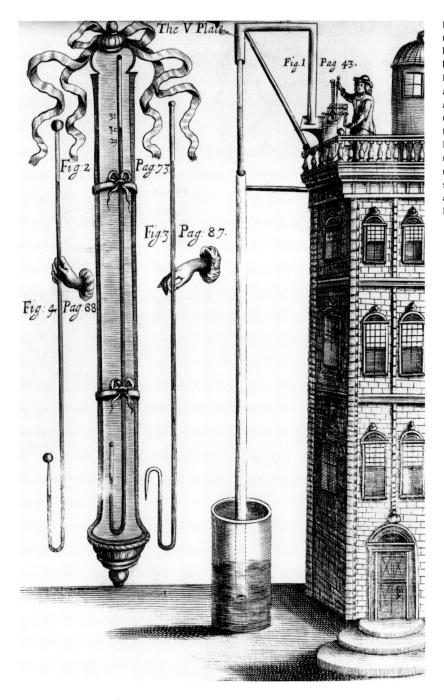

Robert Boyle's experiment to demonstrate the greatest height to which water could be raised by pumping. Boyle stood on a roof approximately 10 metres above a barrel of water and used a pump to suck water from the barrel up a pipe. From 'A continuation of new experiments physico-mechanical, touching the spring and weight of the air, and their effects', by Robert Boyle (1669).

Hooke's design for a vacuum pump was based upon a cylinder with one end closed, containing a piston that stuck out from the open end of the cylinder. The end of the piston was cut with teeth which engaged with a gear wheel that could be wound with a handle to push the piston up, forcing air out through a one-way valve, then pull the piston down, leaving a vacuum in the tube (there is a replica of Hooke's pump in the Science Museum in London). When a glass vessel was attached to the pump via another one-way valve, the piston could be pumped up and down repeatedly, sucking more and more air out of the glass vessel.

At about the time Hooke was developing his pump, in the late 1650s, another Englishman, Richard Towneley, was repeating the experiments carried out by Florin Périer, using a Torricelli barometer, on Pendle Hill in Lancashire. He surmised that the lower pressure of air at higher altitude is because the air is thinner (less dense) there, and mentioned this idea, which became known as Towneley's Hypothesis, to Boyle. Boyle was intrigued, and gave Hooke the task of carrying out experiments to test the hypothesis.

The simplest of these experiments did not involve the air pump. Hooke took a glass tube shaped like the letter J, with the top open and the short end closed. He poured mercury into the tube to fill the U-bend at the bottom (just like the U-bend in a kitchen sink), sealing off the air trapped in the short arm of the J. When the mercury was at the same level on both sides of the U-bend, it meant that the trapped air was at atmospheric pressure. But when more mercury was poured in to the tube, because of its extra weight the pressure increased and forced the air in the closed end into a smaller space. Boyle was not a great one for calculations, but Hooke was, and he made careful measurements of the amount of mercury being added and the amount by which the trapped air was squeezed, which showed that the volume of the trapped air was inversely proportional to the pressure. In other words, if the pressure doubles, the volume is halved; if the pressure triples, the air is squeezed into a third of its original volume, and so on.

Other experiments carried out by Boyle and Hooke did use the air pump, and showed, for example, that water boils at a lower temperature when the air pressure is reduced (which explains why it is hard to make a good cup of tea on top of a mountain). This was a very tricky experiment, as it involved placing a mercury barometer inside a sealed glass vessel where the water was being heated, to monitor the pressure as air was pumped out.

The experimental results were first announced to the world in Boyle's book, *New Experiments Physico-Mechanical Touching the Spring of the Air*, published in 1660. But at that time he did not explicitly spell out the inverse law relating volume and pressure. That appeared in the second edition of his book, published in 1662, and as a result it became known as Boyle's Law, even though Hooke had done the experiments and made the calculations on which the law was based.

All of this was important to scientific thinking, because it supported the idea that the air is made of atoms and molecules, flying around and colliding with one another. It was also important in practical terms, because the realization that air has weight, and that it can be extracted using pistons to leave a vacuum, fed directly into the idea of the steam engine (see page 48).

Robert Hooke may have missed out on getting his name attached to 'Boyle's' Law, but he soon achieved an even greater experimental success as a pioneer of the use of the microscope. In the second half of the seventeenth century other experimenters also studied the world of the very small using lenses to magnify tiny objects, but it was Hooke who did the most thorough job, and explained his discoveries in a book, *Micrographia*, which was published in 1665. It was written in English, unusually for the time (most learned tomes were written in Latin), and easy for any educated person to understand. Samuel Pepys called it 'the most ingenious book that ever I read in my life'.

There were actually two ways to achieve the kind of magnification needed to make microscopy worthwhile. One was pioneered by Hooke's contemporary, the Dutch draper and amateur scientist, Antoni van Leeuwenhoek. His 'microscopes' were single tiny lenses, some no bigger than a pinhead, mounted in strips of metal. They had to be held very close to the eye, and acted as powerful

Seventeenth-century drawing of a flea observed through a microscope by Robert Hooke (1635–1703).

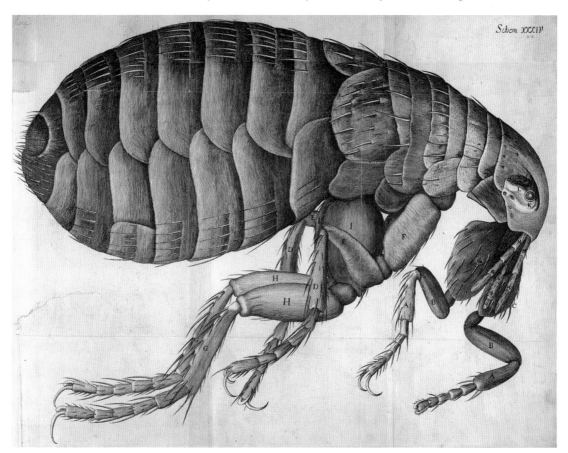

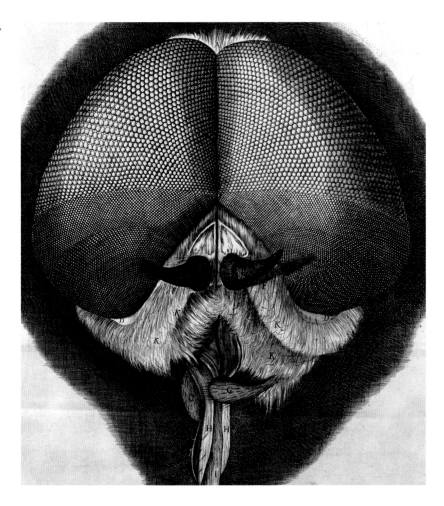

magnifying glasses – so powerful that they could enlarge images 200 or 300 times. These were very difficult to work with, but van Leeuwenhoek made many important discoveries, including tiny living creatures in droplets of water. Hooke used this kind of lens when he needed to pick out the tiniest details in the objects he studied, but he also used a different experimental setup, the forerunner of modern microscopes. These microscopes were made from combinations of lenses, mounted in tubes 6 or 7 inches long (approximately 15–18 centimetres), similar to the way in which telescopes are made. They were easier to work with, but did not give such powerful magnification as the tiny single lenses.

But 'easier' does not mean 'easy'. Anyone who has used a microscope knows that it is important to have a bright light shining on the object being studied at the focus of the instrument. But there was no electric light (or even gaslight) in the 1660s. Candles were not bright enough to do the job. So Hooke used an ingenious arrangement of glass lenses and round containers filled with water to act as spherical lenses to focus light from the Sun (or sometimes a candle) on the object of interest. This worked well for inanimate objects. But he also wanted to study living things, such as ants. These would soon crawl out of the focus of the microscope, but if he killed them then their bodies shrivelled up. He tried sticking them down with wax or glue, but they still wriggled about too much to be studied. Then he hit on the idea of dosing them with brandy to make them unconscious. The ant, he said, was soon 'dead drunk, so that he became moveless'.

Of course, there was no way to photograph the objects under the microscope, and Hooke's book is full of beautiful, detailed drawings of what he saw. A seventeenth-century experimenter had to be an artist as well as a scientist. Hooke showed his readers how much irregularity there is in something as seemingly perfect as the point of a needle or the edge of a razor, and he also showed them how much regularity there is in crystals. This, he said, must result from a regular arrangement of the particles making up crystals – an early hint at the existence of atoms.

But perhaps his most spectacular realization was that fossils are the remains of once-living creatures. In the middle of the seventeenth century, it was widely accepted that these peculiar stones, resembling living things, are just pieces of rock that have been distorted by some unknown process to look like living things. But Hooke, with the evidence of his microscopic studies in front of him, said that fossils are not simply contorted pieces of rock. The details matched the patterns of living things too precisely for that to be true. He said that the fossils we now call ammonites must be 'the shells of certain Shellfishes, which, either by some Deluge, Inundation, earthquake, or some such other means, came to be thrown to that place, and there to be filled with some kind of mud or clay, or petrifying water'. And he realized that because such remains are now found far from the sea, there must have been major changes to the Earth in the past. The reference to a 'Deluge' seems to have been a sop to those who believed in the literal truth of the story of the Biblical Flood. Hooke made his own views clear in a lecture at Gresham College in London, where he said that 'parts which have been sea are now land', and 'mountains have been turned into plains, and plains into mountains, and the like.' A profound inference to draw from looking at tiny objects through a microscope.

Nº. 11 ALL THE COLOURS OF THE RAINBOW

Isaac Newton is widely regarded as the greatest scientific thinker that ever lived, and the emphasis in this appreciation is usually on his skills as a theorist, propounding the laws of motion and, most famously of all, the law of gravity. But like his contemporaries, Newton was also a 'hands-on' scientist – a practical man who did his own experiments, often using equipment he had designed and

Isaac Newton portrayed in front of his own drawing (colour added) showing the splitting of white light from the sun into the spectrum. The illustration also shows that a second prism refracts the original colours, in this case red, without further change once split, or dispersed, the second prism refracts the original colours, in this case red, without further change.

built himself. This approach transformed science in the 1660s, largely because of the influence of the Royal Society, founded in the early years of that decade (it received its Royal Charter in 1663). The motto of the Society was (and is) *Nullius in Verba*, which can be loosely translated as 'take nobody's word for it'. From the outset, they did not simply accept hearsay reports of scientific discoveries, but carried out experiments and demonstrations themselves to test such claims. (In the early days, Hooke was the man who did the experiments.) Newton came to the attention of the Society in 1671 because of his practical skills – he had designed and built a new kind of telescope, useful for astronomical investigations, which focused light using a curved mirror rather than a lens. The telescope was shown to the Royal Society by Isaac Barrow, a Cambridge mathematician who had been one of the first people to recognize Newton's ability. Newton himself was by then Lucasian Professor of Mathematics in Cambridge, but lived a quiet life and largely kept his many discoveries to himself.

The Society was sufficiently impressed to elect Newton as a Fellow on 11 January 1672, and to ask him what else he had been working on. His reply took the form of a long letter (what we would now call a scientific paper) in which he explained his ideas about light and the experiments on which those ideas were based. Newton's key insight was that 'pure' white light is actually a mixture of all the colours of the rainbow. To the ancients, white light represented a pure entity, in the same way that spheres were thought to be perfect. To suggest that white light was a mixture of colours would have seemed to them as ludicrous as the idea that planets did not move in perfect circles in their orbits.

Of course, it was well known that when sunlight passed through triangular prisms or other pieces of glass it produced rainbow patterns of colour. But before Newton people assumed that the white light passing through the glass had become adulterated as it picked up imperfections from the glass, and this had changed its nature. Newton's genius was to devise a simple experiment which proved that this was wrong.

In the first stage of the experiment, he worked in a darkened room with heavy curtains to shut out the sunlight. A small hole in the curtain let through a single beam of light, which fell upon a triangular prism. After passing through the prism, the light shone upon a white wall on the other side of the room, where it was spread out into a rainbow pattern of colours. It was Newton who identified seven different colours in this pattern – red, orange, yellow, green, blue, indigo, and violet.

This could still be explained as a change brought about by the passage of the light through the glass. But in the second part of the experiment Newton put a second prism, reversed compared with the first prism, between the first prism and the white wall. The first prism had spread white light out into seven colours; the second prism did not make the colours even brighter (more 'impure') but rather combined the seven colours back into a single spot of white light. He had turned a 'rainbow' back into white light. The explanation was that white light is really a mixture of all the colours of the rainbow, and that light is bent as it passes through the prisms (or, indeed, inside raindrops). Some colours are bent more than others, so they get spread out or squeezed back together depending on how the prism is oriented. It is as easy to bend them back into a single beam as it is to bend the different colours out of a single beam. This was the first step towards an understanding of spectroscopy (see page 117), which would one day make it possible to determine the composition of the stars, using reflecting telescopes, some of them developed from Newton's design.

Replica of Newton's reflecting telescope.

This was just one of Newton's insights into the nature of light, which were eventually gathered in a great book, *Opticks*, published in 1704, the year after he had become President of the Royal Society. In that book, he summed up his own understanding of the scientific method: 'Analysis consists in making Experiments and Observations, and in drawing general Conclusions from them by Induction, and admitting of no Objections against the Conclusions but such as are taken from Experiment, or other certain Truths.'

THE SPEED OF LIGHT IS FINITE

Newton's remark about experiments 'and Observations' is important. Sometimes Nature does the 'experiment' for us, and the role of the scientist is 'only' to observe what is going on and work out why it has happened. But it often takes a very clever person to work this out. Ole Rømer's discovery of the finite speed of light is a case in point.

In the seventeenth century, there was a great deal of interest in using studies of the eclipses of the moons of Jupiter (discovered by Galileo) as 'clocks' to determine longitude. These eclipses occur at regular intervals, as the moons orbit Jupiter in the same way that the Earth orbits the Sun. The moment when a particular moon disappeared behind Jupiter could be observed from different places on Earth, and the time at which it occurred could be compared with the local time measured from noon. This told the observers how far east or west they were from some chosen reference point. This technique was pioneered by the Italian Giovanni Cassini, who moved to the Paris Observatory in 1671. He sent the Frenchman Jean Picard to Denmark to use observations of Jupiter's moons to establish the exact longitude of Tycho Brahe's old observatory, so that Tycho's records could be tied in to observations from Paris. Picard was helped by a young assistant, the Dane, Ole Rømer, who then went to Paris to work with Cassini (he was also for a time the tutor of the French Crown Prince, the Dauphin).

Over the next few years, Rømer continued to monitor the eclipses of Jupiter's moons, and noticed that these did not always occur exactly when they were expected. He made a particular study of the moon Io, and noticed that the time between one eclipse and the next got shorter when the Earth was moving towards its closest to Jupiter (which happens when it is on the same side of the Sun as Jupiter), and longer when it was moving further away. Cassini himself thought for a time that this might be because light travels at a finite speed. When the Earth is moving towards Jupiter, the time between successive eclipses is shorter because in the time between eclipses the Earth has moved closer to Jupiter, so light from the second eclipse does not have so far to travel and gets here quicker. Similarly, when the Earth is moving away from Jupiter, the light from the second eclipse has further to travel, because the Earth has moved on in its orbit, and takes longer to reach us.

Curiously, Cassini abandoned the idea. But Rømer took it up, and made detailed observations and calculations to develop it further. In August 1676, Cassini, then still enamoured of the idea, announced to the French Academy of Sciences that the official tables of the eclipses of Io, used in calculating longitude, would have to be revised 'due to light taking some time to reach us from the satellite; light seems to take about ten to eleven minutes [to cross] a

Danish astronomer
Ole Rømer (1644–1710)
with the tools of his trade.

distance equal to the half-diameter of the terrestrial orbit.' Cassini also
predicted that the emergence of Io from eclipse on 16 November 1676 would
be seen about ten minutes later than would have been calculated by the
previous method. In fact, the emergence that occurred on 9 November was
observed and matched the new calculations, prompting Rømer to make a more
detailed presentation to the Academy.

Unfortunately, most of Rømer's papers were lost in a fire in 1728, and the
only account we have of that presentation is a rather garbled news story,
which was translated into English and published in 1677 in the Philosophical
Transactions of the Royal Society. But a later document survived and spells

out his calculation and the dramatic result that he announced. With the best estimate available to him for the size of the Earth's orbit, Rømer calculated that the speed of light must be (in modern units) 225,000 kilometres per second. If we make the same calculation, using Rømer's observations, and plug in the modern value for the diameter of the Earth's orbit, we get a speed of 298,000 kilometres per second. This is remarkably close to the best modern measurement of the speed of light: 299,792 kilometres per second.

Although not everyone was convinced at the time, the discovery made Rømer's reputation. He visited England, where he was warmly received and discussed his observations and their implications with such people as Isaac Newton, Edmond Halley and the Astronomer Royal John Flamsteed. They were convinced. In his book, *Opticks*, Newton mentioned that light takes 'seven or eight minutes' to travel from the Sun to Earth. Rømer returned to Denmark in 1681 to become Astronomer Royal and Director of the Royal Observatory in Copenhagen, the spiritual heir to Tycho.

Nº 13   VITAMIN AT SEA

There is a particular kind of experiment that is important in everyday life, although it is not usually called an 'experiment', perhaps for fear of worrying the people taking part. This is the 'medical trial' – a kind of controlled experiment. One of the best examples of a medical trial is also one of the earliest, dating from the 1740s, when James Lind, a surgeon in the Royal Navy, carried out experiments to find a cure for scurvy. The story of scurvy also shows the value of careful experiments and observations, even when it takes a long time for these observations to be explained.

Scurvy is an illness that starts out with a general feeling of being under the weather and lethargy, then develops with spots on the skin, softening and bleeding of the gums, loss of teeth, open wounds on the skin, and eventually death. We now know that it is caused by a lack of vitamin C, but in the eighteenth century nobody knew anything about vitamins. What they

The mouth of a person suffering from scurvy, showing swollen and bleeding gums.

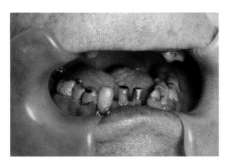

did know was that scurvy was prevalent among sailors and soldiers on a restricted diet of dried meats and grains. The problem was highlighted by the fate of the first British circumnavigation of the globe, by a fleet led by Sir George Anson between 1740 and 1744. Out of an initial complement of about 2,000 men, more than half died from scurvy on the voyage. For a growing naval power such as Britain, finding a treatment or preventative for scurvy was a pressing problem.

Lind was not the first person to suggest that citrus fruits might be used to treat scurvy, but he was the first person to carry out a scientific experiment to test the idea. His ideas about the cause of scurvy were completely wrong – he thought it was caused by a 'putrefaction' of the body, which he hoped to treat with acids. On a voyage in 1747, he tested this idea by adding different acids as supplements to the diet of different groups of men afflicted with scurvy. All the sailors ate the same foods, but every day one group had a quart of cider each, one group had 25 drops of elixir of vitriol (sulphuric acid) added to their diet, one group took six spoonfuls of vinegar each, one group had to drink half a pint of seawater, the members of another group each ate two oranges and one lemon a day, and the last group drank barley water and ate a spicy paste. The seawater drinkers were a 'control', because they were not given any medicine. Hence the term 'controlled experiment'. By the time the experiment ended (because the ship had run out of fruit) the health of the sailors given oranges and lemons had improved dramatically; among the other groups, only the cider drinkers showed a slight improvement.

James Lind (1716–1794).

The Royal Navy took note of the discovery, and some captains began to implement a policy of providing ships with a syrup made from oranges, and also with sauerkraut, which had also proved an effective antiscorbutic (from the Latin term for scurvy, *scorbutus*). Lind left the service soon after this voyage, and although he wrote a book, *A Treatise of the Scurvy*, which was published in 1753, it was largely ignored. But a second 'experiment' was carried out by James Cook on his first circumnavigation of the globe, starting in 1768. His ship carried three tons of sauerkraut. This tasted vile, but Cook persuaded his crew to eat it by a 'method I never once knew to fail with seamen'. He had the food served at first only to officers, who ate it with the appearance of delight. The men soon petitioned for it to be added to their rations, and the result was that there were hardly any incidences of scurvy.

Although shore-based doctors continued to ignore the evidence, experience had shown the Navy what worked, even if they did not know why it worked. In 1794, lemon juice was issued to sailors on board the *Suffolk* on a 23-week voyage to India during which nobody died from scurvy. A year later, lemon juice began to be issued to every ship. The juice proved quite palatable, since it was drunk mixed with the sailors' 'grog', a ration of rum diluted with water. This was later replaced by lime juice, which proved even more effective, and, by the middle of the nineteenth century, American sailors had begun referring to their Royal Navy counterparts as 'lime-juicers' – later shortened to 'limeys' and applied to anyone from Britain.

It was not until the early 1930s that the active ingredient was identified and named ascorbic acid, or vitamin C. Most animals can make their own vitamin C, but monkeys and apes (including humans), guinea pigs and bats are among the few who cannot do so and have to get vitamin C through their diet.

CONDUCTING THE LIGHTNING

In the middle of the eighteenth century, there was no way to generate electric currents (see page 61), but scientists were familiar with the 'static' electricity that can be made by friction, for example by rubbing a glass rod with a silk cloth. This is the same kind of electricity that crackles on a dry day when you pull up a sweater made of synthetic material, and a rod charged up in this way will produce a spark when it is touched to another object. These sparks are like miniature lightning bolts, which led people to speculate that lightning might be a form of electricity. The person who took up the challenge of proving this was the American savant Benjamin Franklin. In 1746 Franklin acquired a glass rod from Peter Collinson, a merchant with an interest in science, who was a Fellow of the Royal Society. He carried out a series of experiments with it.

These experiments convinced Franklin that storm clouds must be electrically charged, like a glass rod rubbed with silk, and that lightning occurred when this static electricity was discharged to the ground, like the sparks that flew when the rod came near another object. But how could this idea be tested? Franklin wrote about his experiments in letters he sent to Collinson in England, and suggested that a tall metal rod, or spike, might be erected during a thunderstorm to draw off electricity from the clouds. The idea was not to encourage lightning to strike the metal rod, but to draw the electricity off gently and capture it in a special kind of glass bottle, known as a Leyden (or Leiden) Jar. Apart from Collinson, nobody at the Royal Society was enthusiastic about Franklin's ideas. But Collinson had them published in the early 1750s, and they were widely read by scientists in mainland Europe.

One of those scientists, Thomas-François d'Alibard, decided to put Franklin's idea to the test. In May 1752 he set up a 40-foot (12-metre) high metal spike in a garden in Marly-la-Ville, in northern France, and drew off sparks from it during a storm. The rod was not, however, struck by lightning, which probably would have killed d'Alibard. Before news of this success could reach America, Franklin had carried out his own version of the experiment in Philadelphia, by flying a kite during a thunderstorm in June 1752.

Like d'Alibard, Franklin knew that it would be dangerous to be struck by lightning, and was simply trying to draw off some of the electric charge from the clouds, along the wet string of the kite to a key attached to the string. To encourage this, a pointed wire was attached to the kite itself, but as a safety precaution he held the kite by a silk ribbon, attached to the string of the kite, as an insulator. Sure enough, electricity was drawn down to the key, and, when an object was moved close to the key, sparks flew across the gap. Franklin even let the sparks leap across, painfully, to his own knuckles.

Peter Collinson (1694–1768).

French scientist
Thomas-François d'Alibard
(1709–1799) carrying out
his lightning experiment
on 10 May 1752, at Marly,
France.

Soon after he had carried out his own experiment, Franklin heard of
d'Alibard's success in France. In October 1752, he wrote to Collinson with
directions on how to repeat his own experiment: 'When rain has wet the kite
twine so that it can conduct the electric fire freely, you will find it streams out
plentifully from the key at the approach of your knuckle, and with this key a
phial, or Leiden jar, may be charged: and from electric fire thus obtained spirits
may be kindled, and all other electric experiments [may be] performed which
are usually done by the help of a rubber glass globe or tube; and therefore the
sameness of the electrical matter with that of lightening completely
demonstrated.'[6]

Other experimenters were not so careful – or rather, not so lucky – as Franklin, and several people were killed by lightning while trying to copy what he had done. In Franklin's case, the electricity was drawn down gradually from the clouds, but he was wrong to think that lightning itself could not strike via the kite. The same flawed thinking was behind his invention of the lightning conductor, in the form of a metal rod attached to the highest point of a building and connected to the earth by a wire. He thought that such a rod would draw off electricity gradually, and prevent a lightning strike. In fact, such a lightning conductor encourages the lightning to strike, but protects the building by acting as a direct route to earth for the lightning, which strikes the metal rod rather than the building itself. But either way, it does work! And all of this proved that lightning is indeed the same phenomenon as static electricity, but on a larger scale.

Nº 15 THE HEAT OF ICE

Ice has an intriguing property, which fascinated scientists studying the nature of heat in the eighteenth century. As well as being intrinsically interesting, these studies had practical implications; it was just at the time steam power was beginning to be harnessed to drive the Industrial Revolution. The curious property is that when ice at the freezing point (0 °C, or, in the units used in Britain then, 32 °F) is heated, its temperature stays the same until all the ice has melted into water. Only then does the temperature of the water increase as more heat is applied. The same sort of thing, of course, happens when other substances, such as metals, are melted, but ice is much easier to study.

Other people had thought that if a lump of ice at the melting point were heated by a tiny amount it would all melt. But the person who studied what was really going on in a careful series of experiments in the 1760s was a professor at Glasgow University, Joseph Black. Whenever Black did experiments, he measured everything that could be measured, as accurately as possible. He had made his name by studying the amount of gas produced or absorbed in chemical reactions. In one of his experiments, a carefully weighed amount of limestone was heated, to produce quicklime, which was then weighed. The quicklime weighed less, because the gas we now call carbon dioxide had been driven off. A weighed amount of water was added to the quicklime to produce slaked lime, which was weighed. Then, a weighed amount of a mild alkali was added to convert the slaked lime back into what weighing proved to be the same amount of limestone that he had started with. Along the way, the differences in weight told him how much gas had been lost or absorbed at each stage. This was quantitative science, as opposed to

Joseph Black (1728–1799).

qualitative science, in which the changes in the character of the substances (their quality) was noted, but there were no measurements of how much they had changed (the quantity).

Black carried over this quantitative approach – a cornerstone of modern experimental science – into his studies of heat. He found that the amount of heat needed to melt a certain amount of ice at 32 °F into water at the same temperature was enough to raise the temperature of the water from 32 °F all the way to 140 °F (or 60° C). He also studied the way water turns into steam, showing that when a mixture of water and water vapour at the boiling point (212 °F, or 100 °C) is heated, the temperature does not increase until all the water has been turned into vapour. And if a certain weight of water – say, a pound – at 32 °F is added to the same quantity of water at 212 °F, the resulting liquid has a temperature of 122 °F (or 50° C), halfway between boiling and freezing. This led him to the idea of 'specific heat', which is the amount of heat required to raise the temperature of a certain amount of stuff by one degree (in modern units, the heat required to raise the temperature of 1 gram by 1 °C). Black coined the term 'specific heat', and also gave the name 'latent heat' to the heat absorbed by a melting substance. And when a liquid such as water freezes, the same amount of latent heat is released as it does so; similarly, latent heat is

Joseph Black giving a practical demonstration of latent heat to students of Glasgow University in the 1760s.

released when vapour condenses into liquid. In Black's own words: 'I, therefore, set seriously about making experiments, conformable to the suspicion that I entertained concerning the boiling of fluids ... I imagined that, during the boiling, heat is absorbed by the water, and enters into the composition of the vapour produced from it, in the same manner as it is absorbed by ice in melting, and enters into the composition of the produced water. And, as the ostensible effect of the heat, in this last case, consists, not in warming the surrounding bodies, but in rendering the ice fluid; so, in the case of boiling, the heat absorbed does not warm surrounding bodies, but converts the water into vapour. In both cases, considered as the cause of warmth, we do not perceive its presence: it is concealed, or latent, and I give it the name of LATENT HEAT.'[7]

These discoveries were noted by a certain James Watt, an instrument maker at the university, who built experimental apparatus for Black, and who went on to develop steam engines.

N⁰· 16   STEAMING AHEAD

There is no clear distinction between experiment and invention. Inventors have to experiment to find out what works, and experimenters often have to be inventors, as the example of the development of the vacuum pump for use in scientific investigation highlights (see page 32). Although this book concentrates on the more obviously experimental end of this spectrum, there is one beautiful example of the synergy between experiment and invention that is so important historically that it simply cannot be overlooked. This is the way in which investigations of the relationship between heat and temperature led to the development of the steam engine, which in turn powered the Industrial Revolution.

In 1763, James Watt was working as an instrument maker in Glasgow, where he became familiar with Black's work, but not, at first, with all of his discoveries concerning latent and specific heat. Watt was asked to repair a scale model of a kind of steam engine developed by Thomas Newcomen, often referred to as an 'atmospheric' engine, because air pressure was just as important in its operation as steam. Such engines had a vertical cylinder, made of metal, containing a metal piston attached at the top (which was open to the air) by a beam to a counterweight. When the space beneath the piston was filled with steam, pressure would increase and the piston would rise. Then, cold water was sprayed into the cylinder, making the steam condense and reducing the pressure so that atmospheric pressure would push the piston down. By repeating this process over and over again, the resulting rocking motion of the beam could be used to drive a pump sucking water out of a mine.

Computer artwork of James
Watt's improved version of
Thomas Newcomen's steam
engine.

By experimenting with the scale model of a Newcomen engine and applying
his understanding of Black's discoveries, Watt realized that this kind of engine is
not very efficient. On every stroke of the engine, the whole cylinder and piston
combination has to be heated up to more than the boiling point of water, in
order to allow it to fill with steam. Then, it has to be cooled sufficiently for the
steam to condense, even though the steam itself (as he later appreciated) gives
up latent heat as it condenses. The heat required to raise the temperature of the
cylinder and piston was thrown away with every stroke of the engine.

Watt realized that it would be much more efficient to have an engine that
used two cylinders, one of which was kept hot all the time and contained the
moving piston, while the other, without a piston, was kept cold all the time.
(He wrote in his journal that this idea came to him on a Sunday afternoon in
May 1765, as he walked across the Glasgow Green.) In his early models, the
cold cylinder, without a piston, was simply immersed in a tank of water. The
two cylinders were connected to each other, but at first outside air was still
used to push the piston down. Steam pushed the piston up as before, but when
the piston reached the top of its stroke a valve opened automatically to let the
steam flow into the cold chamber, where it condensed, reducing the pressure

and allowing the piston to fall. At the bottom of the stroke, another atomatic valve opened to let fresh steam into the cylinder. Soon, this setup was improved by sealing off the piston's cylinder from the atmosphere and using hot steam to push the piston down as well as to push it up. But the key concept was the 'separate condenser'. Watt's steam engine design was patented in1769.

Because Black had not published all of his discoveries, and Watt had a fairly lowly position in Glasgow, at first Watt did not know about Black's work on latent heat, and independently made the same discovery. Specifically, in one series of experiments he noticed that when one part of boiling water is added to thirty parts of cold water, the rise in temperature of the cold water can hardly be measured. However, when the equivalent amount of steam at the temperature of boiling water is bubbled through the cold water, it can raise the temperature of the water to boiling point. This discovery by Watt led to discussions with Black, and Black's understanding of heat helped Watt to make improvements to his steam engine design. Black even helped to fund the development of Watt's idea into a practical machine. But it was in partnership with the venture capitalist Matthew Boulton in the 1770s that Watt developed the engines that drove the Industrial Revolution.

Steam engines under construction at Boulton and Watt's Soho Foundry, at Soho, near Birmingham, UK.

Watt went on to apply science in many other areas of practical importance, including developing a process for bleaching cloth, and a successful early method for copying handwritten letters, a forerunner of the photocopier. In all of this work he provided the archetypal example of an experimenter/inventor, blurring the line between 'pure' science and 'practical' science. As Humphry Davy wrote of him: 'Those who consider James Watt only as a great practical mechanic form a very erroneous idea of his character; he was equally distinguished as a natural philosopher and a chemist, and his inventions demonstrate his profound knowledge of those sciences, and that peculiar characteristic of genius, the union of them for practical application.'[8]

Nᵒ. 17   BREATHING PLANTS AND PURE AIR

In the early 1770s, Joseph Priestley, who was a non-conformist minister, philosopher and scientist, carried out some experiments that hinted at the importance of plants in making air fit to breathe. In 1771, while a minister in Leeds, he put some mint in a pot in a closed container glass with a lit candle. The candle soon went out, but the mint thrived and continued to grow. Twenty-seven days later, without having ever opened the container, he re-lit the candle by focusing sunlight through the glass of the container using a curved mirror. This showed him that the mint had somehow revived the air in the closed container. The following year, he carried out similar experiments with mice. First, he kept a mouse in a similar enclosed container with no plants, and noted how long it took before the mouse collapsed. Then, he repeated the experiment with living plants in the container with the mouse. This time, the mouse survived. Priestley realized that this meant that living plants provided something to the air that animals need in order to live, and that candles need in order to burn. At this time, however, he had no idea what the 'something' was.

In 1774, Priestley left Leeds and was sponsored by Lord Shelburne, who provided him with a base on Shelburne's estate in Calne, Wiltshire. Continuing his experiments there, he studied the gas released by what was then known as the red calx of mercury (now called mercuric oxide) when it was heated by focusing the rays of the Sun on it. He trapped the gas as it was given off, leaving mercury behind, and carried out a long series of experiments with it. He found that a lighted candle put into the gas flared up brightly, and that a glowing taper would re-light if plunged into a tube of this 'pure air', as he called it.

In 1775, he did another mouse experiment. He put a full grown mouse in a container filled with ordinary air, and found that it could survive for only 15 minutes. But when he put a similar mouse in the same container filled with

Engraving of the laboratory of the English chemist, Joseph Priestley (1733–1804).

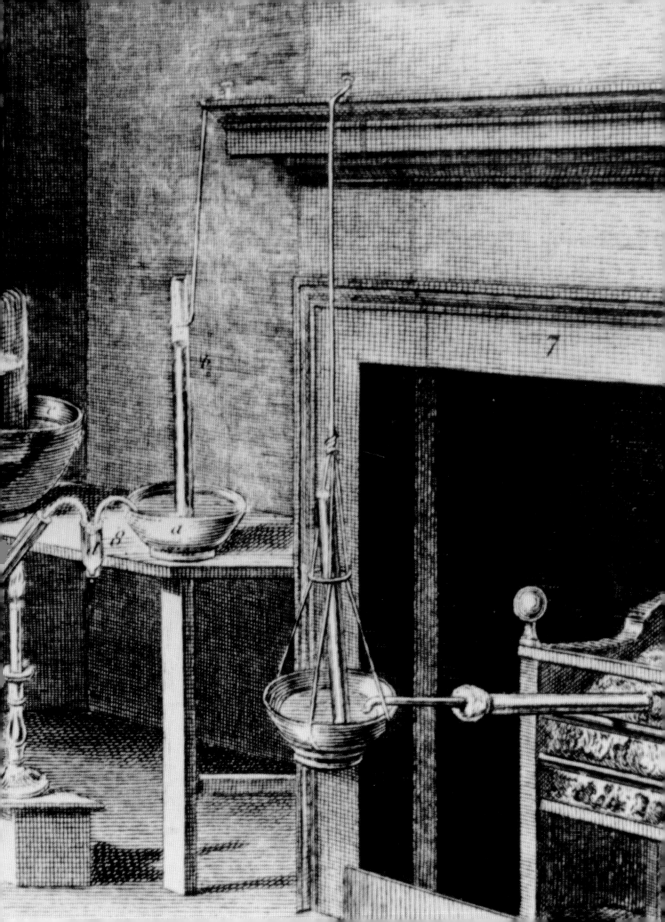

'pure air', it survived for half an hour, and then, when he took the seemingly dead mouse out of the container and warmed it by the fire, it revived. News of these experiments was quickly spread via the Royal Society. Priestley had discovered oxygen, although it would not be given that name until later. A Swedish chemist, Carl Scheele, made the discovery at about the same time, but his results were not published until 1777.

In 1779, a Dutch physician and chemist, Jan Ingenhousz, settled in England after travelling widely in Europe. By then, Priestley had moved on, and Ingenhousz took over his laboratory in Calne, under the same sponsorship. Ingenhousz was also interested in the way air could be 'revived' by plants, and had independently carried out experiments similar to those Priestley had carried out at the beginning of the 1770s. At Calne, he took this work a stage further by putting green plants under water in transparent containers. He observed that bubbles of gas were produced from the underside of the green leaves when they were exposed to sunlight, but that in the absence of sunlight this bubbling stopped.

It was a simple matter to catch the gas produced by the plants and test it. Ingenhousz found that a glowing taper plunged into the gas would relight, and this and other tests showed that it must be Priestley's 'pure air' – what we now call oxygen. As a result of these experiments, Ingenhousz is credited with

Photosynthesis in Canadian pondweed (*Elodea canadensis*). The bubbles around the plant contain oxygen, a by-product of photosynthesis. Photosynthesis is the process by which most plants convert sunlight into chemical energy.

discovering photosynthesis, the chemical process by which plants use energy from sunlight and (among other things) carbon dioxide from the air to build their tissues, with oxygen released as a by-product. Animals use oxygen from the air to power their cells, releasing carbon dioxide as a waste product, so that there is a mutual interdependence between plants and animals. Although these details were worked out only later, the broad picture was clear to Ingenhousz in 1779.

Ingenhousz summed up his discoveries in a book, *Experiments upon Vegetables – Discovering Their Great Power of Purifying the Common Air in the Sunshine and of Injuring it in the Shade and at Night*. He was fascinated by the interdependence between plants and animals, and at the end of the book he wrote: 'If these conjectures were well grounded, it would throw a great deal of new light upon the arrangement of the different parts of the globe and the harmony between all its parts would become more conspicuous.' This comes close to the idea of Gaia, the Earth as a single living organism, two centuries ahead of its time.

Nº· 18   OPENING UP THE SOLAR SYSTEM

The scientific sensation of the 1780s was the discovery of a 'new' planet in the Solar System. This transformed the view of the heavens that had held since ancient times, and began the process of opening up astronomers' images – of, first, the Solar System, and then of the whole Universe.

The Ancients had observed five planets in the sky, named after Roman gods – Mercury, Venus, Mars, Jupiter and Saturn. By the 1780s, it was known that these planets orbit the Sun, with Mercury closest to the Sun and Saturn furthest out, and Earth was also known to be a planet, orbiting the Sun between Venus and Mars. Of course, the planet discovered in 1781, now know as Uranus, was not really new. It had been around for as long as the other planets, orbiting even further out than Saturn, and had even been observed many times, but it had been mistaken for a star or a comet. It is possible that one of the 'stars' identified by Hipparchos in his star catalogue in the second century BC was actually Uranus, although the planet is extremely difficult to spot with the naked eye. Telescopes made it easier to spot, and it is now certain that the planet was identified as a star by John Flamsteed in 1690. A French astronomer, Pierre Lemonnier, observed Uranus several times between 1750 and 1769, without realizing its true nature.

The reason for those missed opportunities, even after the advent of the telescope, is that Uranus is so far from the Sun that it moves very slowly across the sky as seen from Earth. The other planets, as well as being brighter and easier to see, move noticeably against the background stars, which gives them their name, from the Greek word for a 'wanderer'. But this also highlights the importance of applying the 'experimental' method to observations as well as to experiments. It is no good looking at the night sky casually from time to time and speculating about what you see. You have to make methodical observations over a long period of time, keeping careful records and comparing observations from different times to work out what is going on.

That is exactly what William Herschel, assisted by his sister Caroline, was doing in the early 1780s. Herschel was a successful musician, living in Bath, who had developed a passion for astronomy and built his own telescopes, observing the skies from the garden of the house he shared with Caroline. He was actually carrying out a methodical search for double stars when, on 13 March 1781, he noticed an object that appeared in his telescope as a tiny disc, rather than a star-

William Herschel discovered Uranus in 1781 with this telescope.

like point of light. (Stars do not appear as discs even in the best telescopes because they are so much further away than planets.) On 17 March, he looked for the object again, and found that it had moved against the background stars. The natural assumption was that he had found a comet, and he reported the discovery as such to the Royal Society. But when Herschel sent details of his discovery to the Astronomer Royal, Nevil Maskelyne, Maskelyne replied: 'I don't know what to call it. It is as likely to be a regular planet moving in an orbit nearly circular to the sun as a Comet moving in a very eccentric ellipsis. I have not yet seen any coma or tail to it.'

This was a crucial point. Planets move in roughly circular orbits around the Sun, staying at more or less the same distance. Comets dive in from the outer parts of the Solar System, swing past the Sun and head back out into the depths of space. Other observations confirmed Maskelyne's speculation. In particular, the Russian astronomer Anders Johan Lexell calculated the orbit of the object from the available observations and showed that it was indeed nearly circular. In 1783, Herschel wrote to the Royal Society that 'by the observation of the most eminent Astronomers in Europe it appears that the new star, which I had the honour of pointing out to them in March 1781, is a Primary Planet of our Solar System'. By then, he had already been appointed 'King's Astronomer' (not to be confused with Astronomer Royal) by George III, with an income of £200 per annum, which enabled him to become a full-time astronomer.

In order to thank his patron, Herschel named the planet *Georgium Sidus* (George's Star). But this did not go down too well outside the United Kingdom, and the astronomical community eventually settled on the name Uranus – with the stress on the first syllable. Ouranos was, in Greek mythology, the father of Cronus and grandfather of Zeus, who were Saturn and Jupiter in the Roman pantheon, fitting the place of the planet in the Solar System.

ANIMAL HEAT, BUT NO ANIMAL MAGIC

Although earlier experiments, such as those of Priestley and Ingenhousz, had shown the importance of some component of air in maintaining life, at the beginning of the 1780s the details of the process were still far from clear. Like many of his colleagues, the French chemist Antoine Lavoisier, a member of the French Academy of Sciences, speculated that the process resembled a slow form of combustion, with the life-giving component of air being converted into 'fixed air' (carbon dioxide) by, in effect, being burned in the body. But unlike those colleagues, Lavoisier, together with his fellow acadamician Pierre Laplace, carried out a proper scientific experiment, based on quantitative principles, to test the hypothesis.

Their experiment involved a guinea pig, which was placed in a container within another container, itself insulated from the outside world, with the gap between the two containers filled with snow at 0 °C. Under these conditions, the animal was quiet and did not move about much. They waited for ten hours, and collected and measured the water that had melted from the snow as a result of the warmth of the guinea pig's body. It came to 13 ounces (369 grams). Then, in a separate series of experiments Lavoisier and Laplace

Antoine Lavoisier (1743–1794) in his laboratory with his wife and his assistants. His wife (Marie-Anne Pierrette Paulze, 1758–1836) is taking notes at far right.

measured how much fixed air the animal breathed out in ten hours while it was resting. Finally, they compared their guinea-pig measurements with the amount of snow that could be melted by burning enough charcoal to make the same amount of fixed air. This was slightly less than 13 ounces, but the agreement was close enough to convince Lavoisier and a wider circle of scientists that animals keep warm by combining the substance we now call carbon, obtained from their food, with something from the air (oxygen) to make fixed air (carbon dioxide). This was a key step in seeing animals, including human beings, as systems obeying the same laws as burning candles or falling stones.

It was Lavoisier who gave oxygen its name, and who established that burning does indeed involve oxygen from the air combining with the burning substance. This replaced the old idea, still adhered to even by people such as Priestley, that a substance called 'phlogiston' is escaping from the substance as it burns. Lavoisier published his definitive demolition of the phlogiston model in the *Mémoires* of the French Academy in 1786, using the term 'air' where we would say 'gas':

1   There is true combustion, evolution of flame and light, only in so far as the combustible body is surrounded by and in contact with oxygen; combustion cannot take place in any other kind of air or in a vacuum, and burning bodies plunged into either of these are extinguished as if they had been plunged into water.
2   In every combustion there is an absorption of the air in which the combustion takes place; if this air is pure oxygen, it can be completely absorbed, if proper precautions are taken.
3   In every combustion there is an increase in weight in the body that is being burnt, and this increase is exactly equal to the weight of the air that has been absorbed.
4   In every combustion there is an evolution of heat and light.[9]

Lavoisier also gave their modern names to many other substances, and produced the first list of 33 chemical elements, as well as introducing a system of symbols to represent the elements, although not all of them turned out to be elements as we know them today. The key point, though, is that he discarded the old idea of four mystical 'elements' (Earth, Air, Fire, and Water) and replaced it with the idea of an element as a substance that could not be broken down into any simpler substances using chemical processes, while more complex substances were made by combining elements. Indeed, Lavoiser's definition still stands: 'We must admit, as elements, all the substances into which we are capable, by any [chemical] means, to reduce bodies by decomposition.' His naming system used logical rules based on this idea, so that, for example, 'vitriol of Venus' became 'copper sulphate'.

Nineteenth-century artwork of the ice calorimeter developed in the period 1782 to 1784 by the French scientists Antoine Lavoisier (1743–1794) and Pierre-Simon Laplace (1749–1827). The central space (centre right) would contain burning oil (upper right), or an animal such as a guinea pig; the surrounding chamber would contain ice; the outer, melted snow. The lid would be added and the amount of heat produced would be measured in terms of the volume of meltwater from the ice (lower left).

The Calorimeter of Lavoisier and La Place.

His book *Traité Éleméntaire de Chimie* (*Elementary Treatise on Chemistry*) was published in 1789, and laid the foundations of chemistry as a proper scientific subject. It is seen by chemists as their equivalent to Isaac Newton's *Principia*. Lavoisier also spelled out clearly what we now call the law of conservation of mass, which states that matter is neither created nor destroyed in chemical reactions, but only converted from one form into another. In the same year, he also founded a journal, *Annales de Chimie*, which carried research reports about the new science.

As far as such things can be pinned down to a specific year or a specific event, the publication of Lavoisier's book marks the moment when chemistry shed the last traces of alchemy and magic, and became a proper scientific discipline.

Nº. 20 TWITCHING FROGS AND ELECTRIC PILES

D uring the 1790s, a series of experiments led to two major discoveries: that electricity can flow from one place to another, and that electricity is important in operating the muscles of living animals. The second discovery came first, when the Italian physician Luigi Galvani was dissecting a frog. Galvani was also interested in the nature of electricity, and had in his laboratory a hand-cranked machine that could generate electric sparks by the friction of two surfaces rubbing together. This kind of 'static' electricity had been known about since the time of the Ancient Greeks. While Galvani was dissecting a pair of frog's legs, a metal scalpel that had been in contact with the machine and had become electrified touched the sciatic nerve of one of the legs. The leg kicked as if it were still alive.

Galvani carried out many experiments to investigate the phenomenon. He found that legs from a dead frog would twitch if they were connected directly to the electric machine, or if they were laid out on a metal surface during a thunderstorm. But his most important discovery was a result of an observation, rather than a planned experiment.

Luigi Galvani's 1791 experiment on the legs of a frog. The upper diagram shows a silver rod (left) and a brass rod (right) being placed in contact with a foot and the spine of the frog. Bringing the two rods together resulted in the leg twitching as the muscles contracted. The lower diagram shows the metal rod connecting foils of two different metals, with the same result.

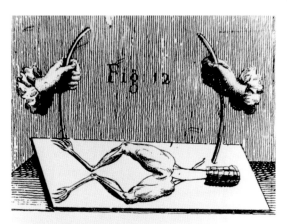

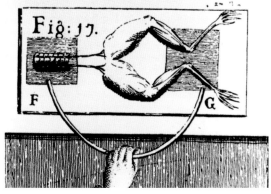

When preparing frogs' legs for study, he would hang them up on brass hooks to dry out. When the hooks came into contact with an iron fence, the legs twitched. In case this was due to some influence from electricity in the air, Galvani took the legs and hooks indoors, away from any source of electricity (including his electrostatic generator) and brought the hooks into contact with iron again. Again, the legs twitched. He concluded that electricity must be manufactured in the body, and stored in the muscles of the frog. He called this 'animal electricity', and proposed that a fluid manufactured in the brain carries this electricity through the nerves of the body to its muscles. But he believed that this animal electricity was something different from the natural electricity of lightning, or the electricity produced artificially through friction.

Most of Galvani's colleagues went along with this idea, which reinforced the idea of a special 'life force', or spirit, which distinguished living things from the non-living world. But one person in particular strongly disagreed. He was another Italian, a physicist called Alessandro Volta. Volta said that electricity was indeed the cause of the twitching of the legs of the dead frogs, but that it had not been stored in the muscles, and that there was no difference between animal electricity and natural electricity. Instead, he suggested that it was being generated from an outside source, an interaction between the two metals, brass and iron, that were in contact with one another.

Volta had already done a lot of work with electricity, including designing and building better friction machines to generate electric charge, and a device to measure electric charge. He first tested his new idea by putting different kinds of

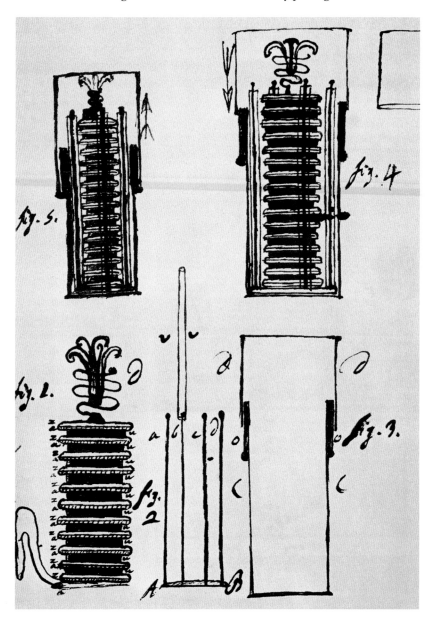

A drawing made by Alessandro Volta (1745–1827) of the first electric battery, called the 'voltaic pile'.

metal in contact with one another and touching the join with his tongue, which tingled as electricity flowed across the join. He realized that the saliva in his mouth was contributing to the effect, and in order to magnify the tiny current he felt with his tongue into something more dramatic he developed a new device, which he described in a letter to the Royal Society in 1800, two years after Galvani had died.

Volta's invention, developed during a long series of experiments, was literally a pile of alternating silver and zinc disks, separated from one another by cardboard discs soaked in brine. When the top disc in the pile was connected to the bottom disc by a wire, electric current flowed through the wire. When there was no connection, no current flowed. The device became known as a 'voltaic pile', the forerunner of the modern battery (the term 'battery' had actually already been used, by Benjamin Franklin, to describe a row of charged glass plates, which he likened to a row of cannon). And it had been developed specifically to disprove Galvani's idea that electric current was part of a life force associated only with living things.

But Galvani and Volta were each partly right. Electricity is generated in the human body, as Galvani thought, but by chemical processes operating in living cells, not solely in the brain. On the other hand, there is nothing special about this electricity, which is exactly the same as the electricity generated in non-living systems such as voltaic piles.

After 1800, scientists could work with electric currents that they could turn on and off as they wished, and they could make the current stronger by adding more discs to the pile, or weaker by taking discs away. Very soon, scientists such as Humphry Davy (see page 85) were using this invention to revolutionize chemistry; one of the first discoveries was that passing an electric current through water decomposes the water into oxygen and hydrogen.

№ 21 WEIGHING THE EARTH

The experiment usually described as 'weighing the Earth' was first carried out at the end of the 1790s, and reported to the Royal Society in 1798. But in fact this experiment was the first determination of the strength of the force of gravity, which turns out to be the weakest of the forces of nature. The experiment was devised by John Michell, who had been a Cambridge professor but gave up his post in 1764 to become a parish priest, although he continued to study science in his spare time.

Michell was the first person to come up with the idea of black holes, in 1783. By then, it was well known that the speed of light is finite (see page 40), and Newton's law of gravity showed that the more massive an object is, the faster you

Model of the gravity
experiment apparatus
used by Henry Cavendish,

have to move to escape from it. Michell calculated that an object with the same density as the Sun but 500 times the diameter of the Sun would be so massive that this 'escape velocity' would exceed the speed of light. He wrote that 'we could have no information from sight' of such an object. It would be, in modern language, a black hole.

He was also interested in more down-to-earth studies of gravity. As well as thinking up an experiment to measure the strength of gravity, Michell got as far as building most of the apparatus needed for the experiment. But he died in 1793, before the experiment could be carried out. All his scientific equipment was left to his old Cambridge College, Queens, but nobody there was up to the task of doing the experiment, so the equipment was passed on to Henry Cavendish, in London, one of the most careful and successful experimenters of his time. Cavendish was a wealthy aristocrat and recluse, who had the money and inclination to devote all his time to the study of science. So the measurement of the strength of gravity became known as the 'Cavendish experiment', even though it was devised by Michell.

The experiment is very easy to understand, but painstakingly difficult to carry out. It was constructed in an outbuilding at Cavendish's house on Clapham Common, then still a village on the outskirts of London. The centrepiece of his experiment was a strong light rod, six feet (1.8 metres) long and made of wood, with a small lead ball on each end. Each of the small balls weighed 1.61 pounds (730 grams). The rod was suspended by a wire from its exact centre, so that it was in balance. Two much heavier lead balls, each weighing 348 pounds (157.85 kilograms), were mounted on swivels so that they could be swung into position at a very precisely measured distance (9 inches, or 22.86 centimetres) from the small balls, each of which had already been weighed very accurately. All of this experimental equipment was placed inside a wooden case so that no air currents could disturb it. Because of the gravitational attraction between the large balls and the small balls, the horizontal bar would try to twist, until it was stopped by the torsion of the wire. By carrying out a long series of experiments, some with no heavy weights and the horizontal bar twisting to and fro like a horizontal pendulum, Cavendish was actually able to measure the force of attraction between each of the small lead balls and a 350 pound weight (158.76 kilograms). This force is about the same as the weight of a single grain of sand. But he also knew the strength of the force of gravity acting between each of the small balls and the Earth itself – this is just the weight of each ball. So from the ratio of these two forces he could work out the mass of the Earth. It is even more impressive that Cavendish completed these experiments not long before his sixty-seventh birthday.

In fact, Cavendish did not give a figure for the mass of the Earth, but quoted a number for its density, which is the mass divided by the volume. On 21 June 1798 he reported to the Royal Society the combined results of a series of eight experiments carried out in August and September 1797, plus nine more carried out in April and May 1798. The figure he reported was 5.48 times the density of water. But he had actually made a slight arithmetical error, and the true density based on his measurements is 5.45 times the density of water. This is very close to the best modern value, which is 5.52 times the density of water. Cavendish was out by just over 1 per cent. And although Cavendish did not do the further calculation himself, such experiments can be used to work out the so-called gravitational constant, G, which is a measure of the strength of gravity.

Nº. 22   BORING EXPERIMENTS ON HEAT

At the end of the eighteenth century, it was widely thought that the phenomenon of heat was associated with a fluid called caloric, which was contained inside objects rather like the way water is contained in a wet sponge. According to this idea, it was caloric flowing out of an object that caused a rise in temperature. A few people disagreed. In particular, the Dutchman Herman Boerhaave had suggested, decades earlier, that heat might be a form of vibration, like sound. But the experiments proving the caloric theory wrong were carried out only in the 1790s, by a colourful American-born scientist then working in Bavaria.

Benjamin Thompson had been on the wrong (British) side in the American war of Independence, as a result of which he became an itinerant scientist/engineer, finding work wherever he could. This led him to Bavaria, where he was so successful in the service of the Elector (Prince) who ruled the region that in 1792 he was ennobled as Graf von Rumford (in English, Count Rumford). Rumford, as he is usually known, became interested in the nature of heat when carrying out experiments to test the power of gunpowder. He noticed that the barrel of a cannon became hotter if it was fired without being loaded than it did if it was loaded with a cannonball, even though the amount of gunpowder used was the same. This is not what you expect from the caloric model, and he had also read about Boerhaave's ideas. One of his roles in Bavaria (at the arsenal in Munich) was to oversee and improve the process of boring out the barrels of cannon, and this gave an opportunity to investigate further.

In making cannon, a cylinder of steel was held in place horizontally, pressed up against a non-rotating drill bit. The steel cylinder itself was rotated, using power from horses plodding round in circles and harnessed to the device by a system of gears, while the drill bit was pushed gradually into the hole being bored. The friction between the drill bit and the steel cylinder soon made the cannon hot. Proponents of the caloric model argued that this was because caloric was being squeezed out of the steel. But Rumford was quick to see the flaw in this argument. The supply of heat seemed to be inexhaustible. As long as the boring continued, heat was produced. If caloric was being squeezed out like water out of a sponge, why didn't it all get used up? Why didn't the 'sponge' dry out?

Using bits of steel left over from the cannon-making which were of no further use, Rumford devised an ingenious experiment. He placed the scrap steel in a wooden box full of water, so that he could measure the amount of heat produced in terms of how long it took to bring the water to the boil (roughly two and a half hours); and he used a dull, discarded drill bit which was little use for boring any more, but which generated a great deal of friction. He found that he could keep on boiling water, refilling the box over and over again, as long as the horses kept walking. The caloric idea disagreed with experiment, so it was wrong. Rumford's results were published by the Royal Society in 1798.

Back in Bavaria, his experiment became something of a party piece, with visitors being amazed at seeing large amounts of cold water made to boil without the use of fire. But Rumford pointed out to them that this was not an efficient way to boil water, because the horses had to be fed hay. If you simply wanted to boil water, it would be much better to do away with the horses and burn the hay itself. With this remark, Rumford was on the brink of understanding the law of conservation of energy, which, echoing the law of conservation of mass (see page 60), says that the total amount of energy is conserved, but that energy can be converted from one form into another. He never quite took that step. But he did write: 'It appears to me to be extremely

American-British physicist Count Rumford (1753–1814), born Benjamin Thompson.

difficult, if not quite impossible, to form any distinct idea of any thing, capable of being excited and communicated in the manner the Heat was excited and communicated in these experiments, except it be MOTION.'

Rumford had no idea exactly what kind of motion might be involved, although he did draw an analogy with the ringing of a bell, but we now know that heat is indeed associated with the motion of atoms and molecules. It was experiments like this that helped to establish the reality of atoms and molecules, as well as inspiring later researchers, such as James Joule (see page 105) in their investigation of heat.

Benjamin Thompson (Count Rumford) demonstrating the heat produced by boring cannon.

THE FIRST VACCINE

T he development of vaccination highlights the importance of applying proper scientific methods, including carefully controlled experiments, to the investigation not just of natural phenomena but to pieces of folk wisdom, which sometimes have a basis of truth. William Gilbert (see page 24) proved that the folklore story that garlic will demagnetise a compass needle was false; but Edward Jenner showed that the folk medicine preventative for smallpox actually worked. He thereby established the scientific basis of vaccination.

Before the development of an effective vaccine, smallpox was one of the main killer diseases. It is actually two related diseases, caused, we now know, by one of two viruses, *Variola major* and *Variola minor* (the name derives from the Latin varius, meaning spotted; victims are covered in small blisters which burst leaving pock marks on the skin). The 'minor' form is less dangerous to life than the 'major' form, but according to Voltaire, writing in 1778, about 60 per cent of the European population caught one or other form of the disease, and a

A doctor peforming a smallpox vaccination on an infant, after a painting of the nineteenth century.

third of the victims died as a result. The World Health Organisation estimated that as recently as 1967 some 15 million people contracted the disease, and two million of them died. As with many infectious diseases, mortality was worse among infants. Because the disease was so prevalent and so serious, before a vaccine was developed people had tried desperate remedies to prevent it. In some parts of the world, people had tried scraping bits of scab from the healing spots of someone with a mild form of smallpox into their skin in the hope that this would somehow make them immune. It seemed to work, and, although some people died as a result of this self-treatment, the practice was brought to England from the Middle East in 1721 by Lady Mary Montagu, the wife of the British Ambassador to Turkey.

Alongside this folk remedy, country lore said that milkmaids were less likely than other people to get smallpox, and this was thought to be linked to the fact that, as a result of their work, they often caught a much milder, but related, illness called cowpox. In the second half of the eighteenth century, several investigators across Europe tested the idea on an ad hoc basis by deliberately infecting people with cowpox. But nobody carried out a proper scientific

A line of villagers waiting to be vaccinated against smallpox and measles during the Smallpox Eradication and Measles Control Program. Photographed in Banso, Cameroon, Africa, in 1969.

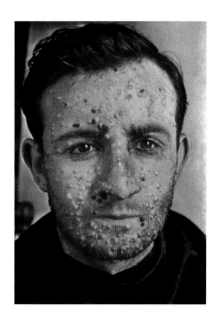

A patient with smallpox, photographed in 1910.

experiment to prove the connection between cowpox and smallpox until Edward Jenner took up the challenge.

Jenner was a country doctor who became completely convinced that catching cowpox provided immunity from smallpox. To prove this, in 1796 he carried out an experiment which seems alarmingly cavalier to modern eyes. He took pus from the blisters of a milkmaid called Sarah Nelmes, and injected this into an eight-year-old boy, James Phipps, who was the son of Jenner's gardener. We do not know what inducement the boy or his father were offered. James developed cowpox, but soon recovered. Jenner then injected the boy with smallpox. He did not develop the disease. Jenner described the experiment in a letter to the Royal Society, but was advised, absolutely correctly, that a single example was not enough to prove his case. So over the following months he carried out a series of similar tests on 23 other people, including his own 11-month-old son. This was enough to persuade the Royal Society, which published the resulting paper in 1798. Jenner coined the word vaccine (from the Latin 'vacca' for cow), and the term later became applied to any use of deliberate injection of a weakened or dead form of a virulent organism to provide immunity from the strong form of the disease.

The medical community was slow to respond to Jenner's discovery, and he gave up his medical practice to concentrate on further research and the promotion of vaccination. Among other things, he developed a method of taking material from human cowpox pocks and drying it onto glass so that it could be taken to where it was needed. In 1853, vaccination against smallpox was made compulsory in Britain, and as the technique spread, together with other efforts to stamp out the disease, smallpox became increasingly rare. In 1801, Jenner had published a pamphlet about vaccination which said, 'the annihilation of the Small Pox, the most dreadful scourge of the human species, must be the final result of this practice'. It took nearly 180 years to achieve that goal, but by the end of the 1970s the World Health Organisation was able to declare that smallpox had been eradicated. The virus now exists only in two government research laboratories, one in the USA and one in Russia, under supposedly secure conditions. Most people think it would be better to destroy these remaining samples.

The important message to take away from this story, however, is not just that Jenner inoculated people with cowpox in order to prevent them catching smallpox. Other people had done that already. The difference is that Jenner then established by experiment (somewhat alarming experiment!) that they were immune to smallpox, and did so repeatedly with enough different subjects to prove his case.

Some people seem to be discovery prone. But this is not a matter of luck; it is a result of careful application of the scientific method of observation and experiment. William Herschel is a case in point. Having made his name in 1781 with careful observations that led to the discovery of Uranus (see page 55), two decades later he made an equally profound discovery through careful experiments here on Earth. This story is less well known than it ought to be, because it is overshadowed by the story of the discovery of Uranus. But what Herschel noticed in 1800 was the beginning of the road that would lead to the revolutionary development of quantum physics a century later (see page 181).

The experiment stemmed from observations. Herschel had been carrying out observations of the Sun, using various coloured glass filters to cut out some of the glare from the light. He noticed that although some of the filters cut out most of the Sun's light, he could still feel the warmth of the Sun coming through the filters. But with other filters, he did not feel much sensation of warmth, even though a lot of light was coming through. In his own words, '[I was using] various combinations of differently-coloured darkening glasses. What appeared remarkable was that when I used some of them, I felt a sensation of heat, though I had but little light; while others gave me much light, with scarce any sensation of heat.'[10]

As a good scientist, Herschel devised an experiment to make quantitative measurements of the effect. It clearly had something to do with the colour of the light, as different filters allowed different colours of light to pass through. So he set up a prism (it was actually a piece of glass he took from a chandelier) to split a beam of sunlight into its different colours and project a spectrum, just as Isaac Newton had done (see page 38). He then took three thermometers, with blackened bulbs to make it easy for them to absorb heat, and placed one on each side of the projected spectrum, to act as 'controls' recording the temperature of the room. The third thermometer could be moved along the spectrum to measure the temperature at different places – that is, the warmth of different colours.

Herschel made a series of observations, each lasting for eight minutes, with the moveable thermometer in either the red, green or violet part of the spectrum. He found that, compared with the control thermometers, after eight minutes the average rise in temperature in the red was 6.9 °F, in the green it was 3.2 °F , and in the violet it was 2 °F. Red light, he concluded, has a stronger heating effect than green light, and green light has a stronger heating effect than violet light. But then he noticed something peculiar. The sunlight that was making the spectrum came through a slit in a screen before reaching the prism, and as the Sun moved across the sky the angle of the beam changed, shifting the projected spectrum. As a result, the central thermometer was left just outside the red end of the spectrum. And it got even hotter. Some invisible form of light seemed to be warming the thermometer.

In further experiments, Herschel moved the thermometer further and further beyond the red end of the spectrum, and found that the heating effect was greatest just beyond the red, but then faded away to nothing. He also tried putting the thermometer beyond the violet end of the spectrum, but could not detect any heating. He had discovered what became known as infrared radiation (although he called them 'calorific rays'), and in further studies he was able to show that it also comes from other sources, not just the Sun – the

A somewhat romanticised and scientifically innacurate portrayal of William Herschel, with his telescope in the background.

warmth you feel from a fire or a hot radiator is infrared radiation. Later studies showed that there is radiation beyond the violet (ultraviolet radiation), although it does not produce this heating effect.

Herschel described his discoveries in a series of papers published by the Royal Society, and suggested that radiant heat and light are both part of the same spectrum – part of which we see and part of which we feel – not two different phenomena, because 'we are not allowed, by the rules of philosophizing, to admit two different causes to explain certain effects, if they may be accounted for by one.' This is a fundamental principle of science, that the simplest explanation is usually best. It is sometimes known as Ockham's Razor after the philosopher-monk William of Ockham, who propounded it in the fourteenth century. Almost immediately after his discovery, powerful support for Herschel's suggestion came from new experimental discoveries concerning the nature of light (see page 78).

N⁰. 25   COSMIC RUBBLE

At the beginning of the nineteenth century astronomers knew of the existence of seven planets. Four of them were relatively small, rocky planets relatively close to the Sun – Mercury, Venus, Earth, and Mars. They are now collectively known as the 'terrestrial' planets. The other three were much larger, and much further out from the Sun – the 'gas giants', Jupiter, Saturn, and Uranus (Neptune would not be discovered until 1846). But there was an intriguingly large gap between the orbit of Mars and the orbit of Jupiter, and many people speculated about a 'missing' planet that might have been expected to fit in to the gap. At the instigation of the Hungarian astronomer, Franz Xaver von Zach (then working in the German city of Gotha), invitations were sent out to twenty-four astronomers asking them to carry out a coordinated search for the 'missing' planet. But one of the people on the list, the Sicilian Giuseppe Piazzi, working in Palermo, found something before the invitation even reached him.

On 1 January 1801, Piazzi was carrying out routine observations when he noticed something that initially looked to him like a star that was not in the catalogues, but which further observations showed to be moving. At first he thought it was a comet, and on 24 January he wrote to a couple of colleagues to announce the discovery. But even at this early stage another thought occurred to him. In those letters, he said 'since its movement is so slow and rather uniform, it has occurred to me several times that it might be something better than a comet'. That 'something better' would, of course, be a planet.

Piazzi fell ill on 11 February, and stopped observing the object, but spread the news with full details of his observations in April 1801. Because of the

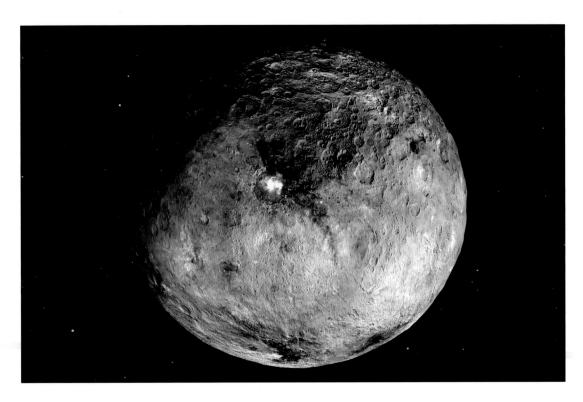

False-colour satellite image of the dwarf planet Ceres, obtained by the Framing Camera on NASA's Dawn space probe.

movement of the Earth in its orbit around the Sun, by that time the supposed planet could not be seen against the glare of the Sun. But there was enough information for the German mathematician Carl Friedrich Gauss to calculate the orbit of the object and predict where it would be found when the Earth had moved further round its orbit. Armed with this prediction, von Zach and another German, Heinrich Olbers, each independently spotted it on 31 December, confirming the orbit and convincing everyone that it was indeed a planet. This was also a neat example of theoretical calculation and experiment or observation coming together to confirm a hypothesis.

As the discoverer of the 'new' planet, Piazzi was able to choose its name. He proposed Ceres Ferdinandea, from the names of the Roman goddess of agriculture and the then king of Sicily. But in an echo of the fate of William Herschel's proposed name for Uranus (see page 55), the 'Ferdinandea', was quickly dropped. Ceres it became. But this was not the end of the story. On 28 March 1802, while looking for Ceres, Olbers discovered a similar object at about the same distance from the Sun. This became known as Pallas. More objects orbiting the Sun in the gap between Mars and Jupiter were soon discovered, and Herschel gave them the name asteroid (meaning 'star like') because: 'They resemble small stars so much as hardly to be distinguished from them. From this, their asteroidal appearance, if I take my name, and call them Asteroids; reserving for myself however the liberty of changing that name, if another, more expressive of their nature, should occur. In all other respects, though, this is a completely inappropriate name.'[11]

At first, the asteroids were thought to be the remnants of a planet that had disintegrated, but detailed studies over many decades suggest that it is more likely that they are pieces of rubble left over from the formation of the Solar System – pieces that failed to make a planet, perhaps because of the gravitational influence of Jupiter, which has ejected much more stuff than we see today from the region. This region of the Solar System where these objects are still found became known as the asteroid belt. There are more than 200 asteroids bigger than 100 kilometres across, and about a million smaller objects. But the total mass is only about 4 per cent of the mass of our Moon; the four largest asteroids, Ceres, Pallas, Vesta, and Hygia, together make up half that mass.

Ceres alone contains a third of the total mass of the asteroid belt, and is a round object, 950 kilometres in diameter, which orbits the Sun once every

Giuseppe Piazzi (1746–1826), the Italian astronomer who discovered Ceres, the largest of the asteroids. The 1000th asteroid to be discovered was named Piazzia in his honour.

4.6 years. Smaller asteroids are irregularly shaped lumps of rock and ice. Because it is big enough to have become round, due to its self-gravity, Ceres was classified as a 'dwarf planet ' by the International Astronomical Union in 2006, but as there has never been a definition of an asteroid except for Herschel's it is also regarded as the largest asteroid.

Nᵒ. 26 FLYING HIGH WITH HYDROGEN

Although the first confirmed human flight in a balloon famously took place on 19 October 1783, when a balloon designed by the Montgolfier brothers (Joseph-Michel and Jacques-Etienne) demonstrated the technique to an astonished crowd in Paris, their balloon used hot air to generate buoyancy. The obvious drawback to this technique is that the balloon has to carry a fire on board, and cannot stay aloft once the fire is exhausted. But even in 1783 it was clear from Boyle's Law (see page 32) and the investigations of gases made by people such as Henry Cavendish and Joseph Black that a balloon filled with hydrogen ought to rise upwards through the atmosphere. This fired the imagination of another Frenchman, Jacques Charles, who designed a hydrogen balloon made of silk, which he coated with rubber to make it gas-tight.

The first small *Charlière* (as such balloons became known) was launched from the Champs de Mars (the present-day site of the Eiffel Tower) on 27 August 1783. It was a 35-cubic-metre capacity sphere capable of lifting about 9 kilograms. After a flight northward covering 21 kilometres and lasting 45 minutes, the balloon landed in the village of Gonesse, where the local peasants, thinking it the work of the Devil, hacked it to bits with knives and pitchforks.

Less than two months after the Montgolfier brothers' first flight, on 1 December 1783, Charles launched a manned hydrogen balloon from the Jardin des Tuileries, carrying himself and Nicolas-Louis Robert, who had devised the method of coating silk with rubber. This hydrogen balloon had a volume of 380 cubic metres, and the ascent was watched by a vast crowd, estimated at 400,000 people – half the population of Paris – and including Benjamin Franklin and Joseph Montgolfier. The balloon reached a height of about 1,800 feet (550 metres) and landed at sunset in Nesles-la-Vallée, 36 kilometres away, after a flight of 2 hours and 5 minutes. Charles and Robert alighted unscathed, just after sunset. Charles then took off again on his own, and reached an altitude of about 10,000 feet (3,000 metres), high enough to see the sun again. But pain in his ears made him release gas and land.

Charles and Robert carried a barometer and a thermometer with them to measure the pressure and the temperature of the air, but these observations were very much a secondary consideration. The first truly scientific balloon ascents were made in 1804, by Joseph Gay-Lussac and Jean-Baptiste Biot, using hydrogen-filled Charlière balloons. Gay-Lussac was thoroughly familiar with the behaviour of gases, and in 1802 had spelled out what is now known as Gay-Lussac's law, which says that if the mass and volume of a gas are kept constant then the gas pressure is proportional to the temperature (on what is now known as the 'absolute' or Kelvin scale, where zero is −273.16 °C). He also discovered that equal volumes of gases expand by the same amount for the same increase in temperature. This is usually known today as 'Charles's law', because Jacques Charles discovered it in the mid1780s, but as he had not published the discovery, Gay-Lussac found it independently.

Joseph-Louis Gay-Lussac (1778–1850) and Jean-Baptiste Biot (1774–1862) making the first balloon ascent specifically for scientific purposes, in 1804.

Gay-Lussac and Biot made an ascent together from the garden of the Conservatoire des Arts et Metiers, on 24 August 1804. The object of the flight was to measure changes in magnetism and the composition and humidity of the air at different altitudes. They reached an altitude of 4,000 metres, and showed that the Earth's magnetic field does not vary noticeably with altitude, but did not achieve any other significant scientific results. So Gay-Lussac obtained a larger balloon and made a solo ascent a few weeks later, on 16 September. This time the balloon took him to an altitude of 7,016 metres (some 23,000 feet; for comparison, the height of Mount Everest is 29,035 feet). This was far higher than anybody had previously flown. At this altitude Gay-Lussac recorded a temperature of –9.5 °C (9.5 °C below freezing), but stayed to make his measurements of the moisture of the air, magnetism, and so on. The magnetic measurements confirmed that, within the accuracy of his instruments, the magnetic field is constant, even up to this altitude. But he noted that he had considerable difficulty in breathing, and that the air was so dry that his mouth and throat became parched to the point where it was painful to swallow a piece of bread. He also recorded an increased pulse rate. But because Gay-Lussac collected samples of air at different altitudes during his flight, he was able to analyse them in the comfort of his laboratory in Paris, rather than struggling to make all the measurements with frozen fingers while in the basket of the balloon. This analysis showed that the composition of the atmosphere does not change with increasing altitude.

Nº. 27   LIGHT IS A WAVE

Isaac Newton had explained the way that light is reflected off mirrors and refracted (bent) by prisms in terms of a stream of tiny particles bouncing off things or being deflected round corners. Although the Dutchman Christiaan Huygens and Newton's contemporary Robert Hooke each developed an alternative explanation, or model, which involved waves, there were no experiments that could test these ideas, and Newton's model dominated scientific thinking for more than a century. Then, the English polymath Thomas Young devised an experiment that proved that light travels as a wave.

Young began to experiment with sound while he was at Cambridge in the 1790s. He appreciated that sound travels as a wave through the air (or through another medium), and he was interested in the phenomenon of interference, when two sound waves affect one another, either adding up to make a loud noise or cancelling each other out to leave a quieter noise. This led him to think about the behaviour of waves in general, and to devise an experiment to test the wave model of light.

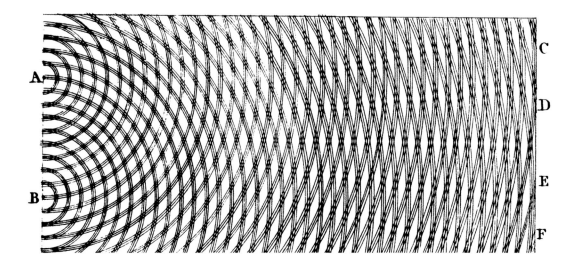

Thomas Young's own illustration of wave interference. From his book, *A Course of Lectures on Natural Philosophy and the Mechanical Arts.*

First, he used an apparatus called a ripple tank to study the behaviour of waves in water. This is just a shallow tank of water in which the ripples can be observed. When Young (or anyone else) put a wall across the tank with a small gap in it, ripples generated on one side of the wall would pass through the hole and spread out on the other side in a semicircular pattern. Then, if a second wall, with two holes, was placed in front of these spreading ripples, ripples would spread out from both holes and interfere with one another, making a pattern with some extra large waves in places where the peaks from each set of ripples combined, and much smaller waves where the peak of one ripple overlapped with the trough of another ripple. This is an interference pattern.

In the early 1800s, Young adapted this experiment to study light, and refined it until the results were unambiguous. In a darkened room, a simple screen of card with either a tiny pinhole or a razor slit was placed in front of a source of light. On the other side of the screen there was another screen with two holes, or slits. And beyond that, there was a blank screen where light that had passed through the whole experiment arrived. The light arriving at this final screen made a pattern of light-and-dark stripes, an interference pattern exactly equivalent to the interference pattern of water waves in the ripple-tank experiment. There could be no doubt that light travels as a wave. This became known as 'Young's double-slit experiment', or more colloquially as 'the experiment with two holes". Part of the power of the experiment is that it is so simple, and yet so profound. As Young put it, when presenting his results to the Royal Society in 1803, 'The experiments I am about to relate … may be repeated with great ease, whenever the sun shines, and without any other apparatus than is at hand to every one."

And yet, there were doubters. Newton was held in such awe that some people refused to accept that he could have been wrong. The idea that you could make a dark stripe by adding two bright beams of light together also

Waves pass through a barrier. Concentric circles of waves emanate from two apertures in a barrier (black line, lower frame). They originate from parallel waves (at bottom) striking the barrier flat on. From each of the two point sources or apertures, the waves spread forming semicircles that fan out. An interference pattern can be seen when concentric circular waves strike and pass over each other.

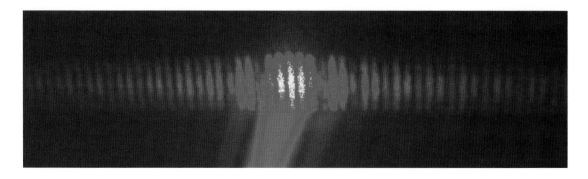

A modern double-slit experiment demonstrates the wave nature of light.

confused them. But a few years later a Frenchman, Augustin Fresnel, proved that this was indeed possible.

This was at the time of the Napoleonic Wars, and Fresnel, in France, seems to have been unaware of Young's work. He independently developed a wave model of light, and wrote it up for a prize competition organized by the French Academy in 1817, asking for explanations for the phenomenon of diffraction, which bends light round corners (by a tiny amount). One of the judges of the competition, Siméon-Denis Poisson, pointed out what he thought was a fatal flaw in Fresnel's model. According to the wave model, if a small round object, such as a piece of lead shot, was placed in front of a beam of light, the light waves would bend round the object to make a bright spot of light on a screen directly behind it, where both 'common sense' and the particle model said the shadow would be darkest. So the judges arranged for an experiment to prove (as they anticipated) that Fresnel was wrong. But the experiment found the bright spot, now known as Poisson's spot, exactly where the wave theory had predicted. This is one of the best examples of a theory being proved right by an experiment carried out in the hope of proving it wrong. Fresnel got his prize, and the wave theory of light was established. Shocking though it might have seemed to some people, even Newton had not always been right.

Nᵒ. 28   DISCOVERING ATOMS

The 'discovery' of atoms in the early years of the nineteenth century resulted from the accumulation of evidence from several different lines of experiment and observation, which were pulled together and formulated by the English scientist John Dalton. Dalton had a lifelong interest in meteorology, and his first steps towards an atomic theory came from the realization that the atmosphere consists of a mixture of gases, with different chemical properties, which do not settle out into different layers but are intermingled. In particular, of special interest to a meteorologist, when water evaporates to make a gas the vapour mingles with the air that already exists, occupying the same space at the same time. It does not push the air out of the way to make a separate layer of water vapour, above or below the air.

This suggested to Dalton that water vapour and air (and other gases) must be made up from separate particles with space between them. You might make an analogy with a box full of pebbles, which has spaces between the pebbles themselves, but no room for more pebbles. But a fine sand, itself made from separate particles, can be poured into the spaces between the pebbles, occupying the same box – that is, the same volume of space. Dalton's experiments showed that the pressure exerted by a certain volume of gas at a certain temperature is always the same as the sum of the pressures that would be exerted by each gas in the mixture on its own. For example, if a container holds a litre of carbon dioxide at atmospheric pressure, and another container holds a litre of nitrogen at atmospheric pressure, if the two gases are squeezed into a single litre volume, the resulting pressure will be twice atmospheric pressure. This is known as Dalton's law of partial pressures, which he spelled out in 1801.

At around the same time, Dalton discovered another law. Although this one does not bear his name, it demonstrates the idea of atoms even more clearly. By heating different gases and measuring their pressure, he found that 'all elastic fluids under the same pressure expand equally by heat'. A quantitative way to measure this is to have a vertical cylinder in which some gas is trapped by a piston, with a weight on the piston applying a steady downward pressure. If the cylinder is heated, the gas expands, pushing the cylinder up even though the pressure is the same. But a simple qualitative demonstration of the same effect can be observed using a child's balloon. If the balloon is inflated to its maximum size at room temperature, its skin will be as tight as that of the proverbial drum. But if the balloon is now cooled with ice water, the pressure of the gas inside will go down, and the skin of the balloon will become flabby and wrinkly. In effect, the gas has shrunk. Let the balloon warm up again, and the skin again becomes tight. An even more dramatic demonstration of the effect can be

Balloon expanding as it warms. The balloon on the left has been cooled to −198 °C using liquid nitrogen. The molecules of air in this balloon have less energy when cooled and move slower, producing less pressure on the inside of the balloon, so it has collapsed to a smaller size. The balloon on the right has been allowed to warm up to room temperature and has regained its normal air volume. In this case the air molecules have more energy and move faster, therefore exerting more pressure on the inside of the balloon, making it expand to a larger size.

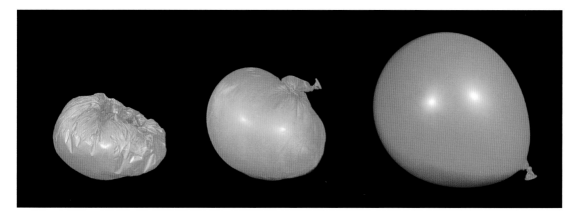

Dalton's Atomic Symbols

SCIENCE: A HISTORY IN 100 EXPERIMENTS

OPPOSITE **John Dalton's
(1766–1844) table of atomic
symbols. Although he
had the right idea about
atomic theory and chemical
reactions, some of his
identifications of elements
and compounds are now
known to be wrong. Lime,
for example, is a compound
of calcium, and water is
not HO.**

observed by cooling the balloon with liquid nitrogen. The same gas at the same
pressure occupies a greater volume at higher temperatures.

The same thing was noticed by the Frenchman Joseph Gay-Lussac about a
year later, but he discovered that it had previously been observed by his
compatriot, Jacques Charles, in 1787. The news had not spread, because Charles
did not publish his discovery. But Gay-Lussac promoted it under the name
'Charles's Law' (see page 77). The obvious explanation is that gases are indeed
made of atoms, and that they are somehow pushed further apart from one
another at higher temperatures.

All of this led Dalton to develop the idea that different gases are made of
different kinds of atoms, with different sizes, different weights. This led him to carry
out chemical experiments comparing the quantity of different elements involved
in various reactions. Among other things, by studying the gases now known as
ethylene (olefiant gas) and methane (carburetted hydrogen), he found that although
both are made only of carbon and hydrogen, methane had exactly twice as much
hydrogen in relation to carbon as ethylene. Such studies led him to a fuller
exposition of the atomic theory, published in 1808, which can be summed up as:

Substances are composed of indivisible particles called atoms.

Each element consists of a characteristic kind of identical atom. Thus there
are as many kinds of atoms as elements. Atoms of a particular element are
perfectly alike.

Atoms are unchangeable.

When elements combine to form a compound, the smallest portion of the
compound is a group containing a definite number of atoms of each element.

In chemical reactions, atoms are never created nor destroyed, but only
rearranged.

This remains the basis of our modern understanding of the structure of
matter, except that atoms are no longer thought to be indivisible (see page 136).

Nº 29   ELECTRIFYING SCIENCE

Soon after Alessandro Volta invented the electric pile (battery), the British
chemist Humphry Davy investigated the relationship between electricity and
chemistry. This is a classic example of the interplay between science and
technology. Scientific investigation led to the invention of the battery, which
provided a source of electric current that was then used in further scientific
investigations. That in turn would lead to new technologies (see page 94), and
so on. As has often proved to be the case, the availability of a new tool meant
that a large number of similar experiments could be carried out to probe the
nature of the material world.

Davy started out from the correct deduction that the electricity produced by a voltaic pile was a result of chemical interactions going on in the pile. So he turned things around by looking at what chemical reactions were stimulated when an electric current was passed through different substances. His key discovery was that in many cases the effect of the current is to break a compound down into its component parts, a process now known as electrolysis. This led to the discovery of 'new' elements, in particular metals.

Davy began studying 'voltaic action' in 1800, when he was based in Bristol, and soon confirmed the link between chemistry and electricity by passing an electric current through water. A wire connected to the positive end of a battery (later known as the anode) was dipped in one end of a tank of water, while a wire connected to the negative end of the battery (later known as the cathode) was dipped in the other end of the tank, so that an electric current flowed through the water. The effect of the current was to produce bubbles of oxygen at the anode and bubbles of hydrogen at the cathode – and, of course, Davy already knew that oxygen gas and hydrogen gas combined to make water. The water had been decomposed by electricity into its constituent parts.

Davy's investigations of this phenomenon had to be put on hold, because in 1801 he moved to a new post at the Royal Institution (itself new) in London, and in 1802 he became Professor of Chemistry there. His duties kept him busy on other projects for a time, but in 1806 he carried out a series of 108 experiments involving electrolysis. The basic experimental setup was improved by attaching the two wires (anode and cathode) to metal plates or rods dipped into the substance being studied. Davy presented the results of his work in a lecture to the Royal Society later that year, and went on in 1807 to electrolyse molten salts. He produced potassium from caustic potash (now known as potassium hydroxide), and sodium from caustic soda (sodium hydroxide). The experiment itself was very simple. Potash is found in the ashes of burned plants, which Davy collected in a small crucible. When an electrical current was passed through the ashes in the same way as in the electrolysis of water, the potash got hot enough to melt, and pure potassium collected around the cathode. These two metals had not previously been differentiated by

Carbon electric arc lamp.

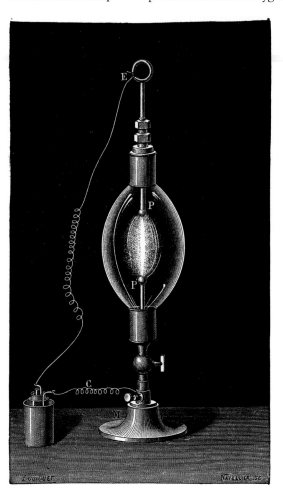

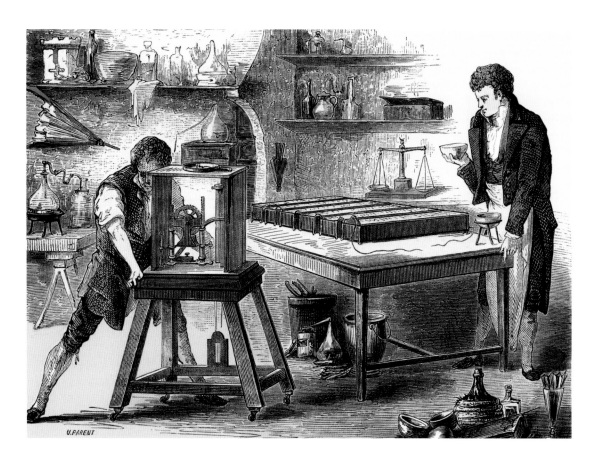

U.PARENT

Nineteenth-century engraving of Humphry Davy (right) using electrolysis to isolate the metallic element in so-called alkalis and alkaline earths.

chemists, but Davy showed that in spite of their similarities they are different elements. All this work was so important that Davy was awarded a French prize of 3,000 francs for the year's best work on galvanic electricity, even though Britain and France were at war.

In further experiments (really, the same experiment repeated with different substances), Davy isolated magnesium, calcium, strontium, and barium. In a slight deviation from this work, in 1809 he made, or discovered, another use for electricity when he took the two wires from a battery and connected them to the opposite ends of a strip of charcoal. The electric current passing through the charcoal heated it until it glowed; Davy had invented the electric light, in the form of an arc lamp. This was essentially the end of his experimental research on electricity, although he did many other things (including isolating and naming the gas chlorine gas). In 1812, his achievements were marked by a high honour – he was knighted. This has more than personal significance, and is a landmark in the history of science. Davy was the first person in Britain to be knighted for his scientific work (Isaac Newton got his knighthood for political reasons), which indicates the increasing importance of science in early nineteenth-century society.

QUANTIFYING CHEMISTRY

B uilding from the work of John Dalton and Humphry Davy, the Swedish chemist Jöns Jacob Berzelius took a major step towards the modern understanding of atoms, molecules and chemistry. Because opposite electrical charges attract one another, he realized that if electrolysis of, for example, water broke it down into hydrogen (attracted to the negatively charged cathode) and oxygen (attracted to the positively charged anode), then water itself must consist of a combination of hydrogen and oxygen atoms held together by some form of electrical attraction – oxygen atoms with an overall negative charge and hydrogen atoms with an overall positive charge (in modern terminology, such charged atoms are called ions).

In order to investigate further, Berzelius carried out a series of experiments in which he measured the proportions of different elements present in various compounds. His declared aim was 'to find the definite and simple proportions in which the constituents of inorganic nature are bound together.' The archetypal example involved taking an oxide of iron and heating it in a stream of hydrogen gas, so that all of the oxygen is extracted and combined with hydrogen to make water; he carried out similar experiments with other metal oxides. By comparing the total weight of the oxide at the start of the experiment with the final weight of iron at the end of the experiment, he knew what weight of oxygen was combined with that weight of iron in the compounds. As well as investigating other metal oxides, Berzelius studied the proportions of oxygen to sulphur in two oxides of sulphur, now known as

The list of chemical symbols for the elements devised by Berzelius.

Element	Berz.	present	Element	Berz.	present	Element	Berz.	present
Aluminium	Al		Hydrargyrum			Potassium	Po	K
Argentum (Silver)	Ag		(Mercury)	Hg (Hy)	Hg	Rhodium	Rh (R)	Rh
Arsenic	As		Hydrogenium	H		Silicium	Si	
Aurum (Gold)	Au		Iridium	I	Ir	Sodium	So	Na
Barium	Ba		Magnesium	Ms	Mg	Stibium		
Bismuth	Bi		Manganese	Ma (Mn)	Mn	(Antimony)	Sb (St)	Sb
Boron	B		Molybdenum	Mo		Strontium	Sr	
Calcium	Ca		Muriatic Radicle			Sulphur	S	
Carbon	C		(Chlorine)	M	Cl	Tellurium	Te	
Cerium	Ce		Nickel	Ni		Tin	Sn (St)	Sn
Chromium	Ch	Cr	Nitric Radicle	N		Titanium	Ti	
Cobalt	Co		Osmium	Os		Tungsten	Tn (W)	W
Columbium	Cl (Cb)	Nb	Oxygenium	O		Uranium	U	
Cuprum (Copper)	Cu		Palladium	Pa	Pd	Yttrium	Y	
Ferrum (Iron)	Fe		Phosphorus	P		Zinc	Zn	
Fluoric Radicle	F		Platinum	Pt		Zirconium	Zr	
Glucinum	Gl	Be	Plumbum (Lead)	Pb (P)	Pb			

sulphur dioxide and sulphur trioxide. And in order to present his results in a clear, easily understood fashion, he invented the system of identifying elements by the initial letter of the name, or Latin name; if there might be confusion between two elements whose names began with the same letter an extra letter was added (for example, carbon became C, so calcium became Ca; Sulphur became S, so silicon became Si).

Jöns Jacob Berzelius (1779–1848).

In order to turn all of this into a system for representing the structure of molecules, Berzelius had to know the relative weights of atoms of different elements. Because hydrogen is the lightest element, he set the weight (or mass) of hydrogen as one unit, and worked out the atomic weights of other elements in proportion. At first, there was some confusion because chemists made the assumption that water consists of equal numbers of hydrogen and oxygen atoms, written as HO in the new notation. But it became clear that everything fitted together better if there were two hydrogen atoms in each molecule of water, so that each hydrogen atom weighed half as much as previously thought. Berzelius would have written the revised formula for water as H^2O; but the convention was soon altered to its modern form, H_2O. These studies also give the oxides of sulphur that Berzelius studied their names, from the proportions of oxygen found in the different molecules – sulphur dioxide is SO_2 and sulphur trioxide is SO_3.

There is one other important difference between the modern chemical conventions and those of Berzelius. Instead of defining the basic unit of weight, or mass (the atomic mass unit), as the mass of a single hydrogen atom, it is now defined as one-twelfth of the mass of an atom of the most commonly occurring form (isotope) of carbon, known as carbon-12. However, this is a subtle distinction that is only really important for chemists and physicists.

In 1818, Berzelius published a table giving the chemical compositions of nearly two thousand compounds, and the atomic weights (actually measured relative to the weight of an oxygen atom in this case) for 45 of the 49 elements that were known at the time. He had determined 39 of these atomic weights himself, with the other six being determined by his students. There were only so many elements to list in this way because in the course of their experiments Berzelius himself and his students had identified many 'new' elements, including cerium, selenium, thorium, lithium, vanadium, and several of the elements known as rare earths. In 1819, he wrote that 'every chemical combination is wholly and solely dependent on two opposing forces, positive and negative electricity, and every chemical compound must be composed of two parts combined by the agency of their electrochemical reaction'.[12]

But even Berzelius still thought that the chemistry of living things was different from the chemistry of non-living things and involved a 'vital force'. He coined the term 'organic compound' to refer to the molecules built around carbon chemistry that seemed to be uniquely involved in biological chemistry, with the rest being termed 'inorganic'. The beginning of the end of this chemical misconception occurred just ten years after Berzelius published his table of atomic weights (see page 88).

Nᵒ. 31 THINKING ABOUT THE POWER OF FIRE

The rapid development of the steam engine following James Watt's experiments (see page 48) was almost entirely a result of trial and error. The engineers had found what worked, but there was no proper theory of *how* such engines worked. The situation was rectified in the 1820s by a Frenchman who took the opposite approach. The engineers had tinkered with machinery to make it work, but Sadi Carnot carried out 'experiments' entirely in his head, and with calculations on paper, to work out what was going on. Such 'thought experiments' proved fundamental to the development of science in later decades.

This was of more than abstract importance. One of the questions that Carnot sought to answer was whether there was any limit to the efficiency of a steam engine, and if so what that limit might be. This was of great practical interest, as it affected the amount of fuel (in those days, coal) that a perfect, or near perfect, 'ideal' engine would consume.

Carnot imagined a perfect steam engine operating a piston in a cylinder, drawing its power from the difference in temperature between a hot 'reservoir' (the fire, in a real steam engine) and a cold 'reservoir', which might in real life be a tank of cold water or even the atmosphere. He described what was going on in a four-step process, now known as a Carnot cycle. First, the gas in the cylinder

Burning fuel in the boiler at Queens Mill in Burnley. The mill is powered by a steam engine, built more than a hundred years ago. The mill is believed to be the only steam-powered weaving mill working in the world. In its heyday the boiler consumed 6 tons of coal a day.

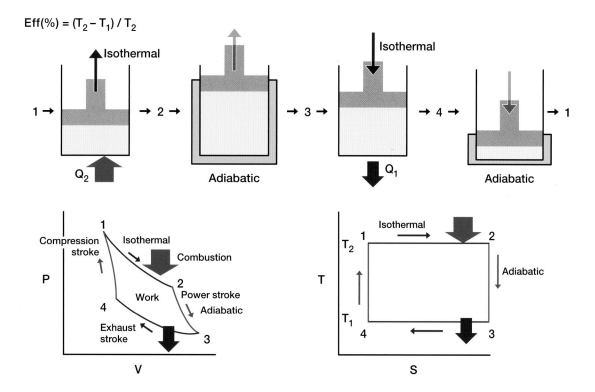

$$Eff(\%) = (T_2 - T_1) / T_2$$

expands at a constant temperature, that of the hot reservoir (isothermal expansion), pushing the piston out. Then, the gas continues to expand while it cools down to the temperature of the cold reservoir (adiabatic expansion). In the third step, the gas is compressed isothermally at the cold temperature, with heat generated by the squeezing flowing out from the gas to the cold reservoir. Finally, the gas is compressed further and heats up as a result, back to the temperature of the hot reservoir. This final step is now known as isentropic compression, although the term was coined only later. At the end of the cycle, everything is back in the state that it started from. But this is only possible because a fire has been providing energy to keep the hot reservoir hot. In effect, the cycler transfers heat from the hot reservoir to the cold reservoir.

Carnot was able to show, by calculating the work done at each stage of the process, that all completely reversible heat engines operating between the same pair of temperatures have the same efficiency (that is, they use the same amount of fuel to do the same amount of work), and that this – the Carnot cycle – is the most efficient way to make use of the temperature difference for any pair of temperatures. A real steam engine, of course, can never be that efficient, because it will lose heat along the way, and suffer from friction. But Carnot showed that no matter how good engineers might get at overcoming these practical difficulties, there is a definite limit to the work available from a heat source, and that replacing the steam with some other working fluid could not alter this limit.

The Carnot cycle, showing how the pressure/volume and temperature/entropy change in the course of the operation of the engine.

Although it might be possible to make engines that are more efficient than steam engines, they could never be more efficient than an ideal machine operating on the Carnot cycle. And he found all this without getting his hands dirty!

A key feature of Carnot's calculations is that a heat engine can be made more efficient if the temperature of its hot reservoir is increased. Decades later, Rudolf Diesel used this realization in his design of the engine that bears his name, which has a 'hot reservoir' much hotter than that of a steam engine.

Carnot published his discoveries in 1824, in a treatise, *Réflexions sur la Puissance Motrice du Feu* (*Reflections on the Motive Power of Fire*), in which he wrote that 'the motive power of heat is independent of the agents employed to realize it; its quantity is fixed solely by the temperatures of the bodies' between which heat is transferred. This was a precursor to the second law of thermodynamics, one of the most important laws in science. It is the second law which quantifies the fact that things wear out, and provides us with a measure of the arrow of time. But the significance of Carnot's work was not appreciated at the time. Partly because Carnot died young (of cholera, in 1832), it was left for Rudolf Clausius and William Thomson (Lord Kelvin) to rediscover these ideas and formalise the study of thermodynamics and the concept of entropy, which is a measure of the amount of disorder in the Universe. The next steps in that direction would be taken by Julius Mayer and (separately) James Joule (see page 105).

Nᴼ. 32   A RANDOM WALK

xperiment and observation came together in a series of studies which between them proved the existence of atoms. The date of the key experiment was 1827, but the observations went back to Roman times, and the explanation of what was going on was not completed until the twentieth century.

We have all seen motes of dust dancing in a beam of sunlight. Some of this motion is caused by convection and other currents in the air. But the Roman Lucretius spotted that something else is going on. Around the year 60 BC, in *De Rerum Natura* (*On the Nature of Things*), he wrote: 'Observe what happens when sunbeams are admitted into a building and shed light on its shadowy places. You will see a multitude of tiny particles mingling in a multitude of ways … their dancing is an actual indication of underlying movements of matter that are hidden from our sight … It originates with the atoms which move of themselves. Then those small compound bodies that are least removed from the impetus of the atoms are set in motion by the impact of their invisible blows and in turn cannon against slightly larger bodies. So the movement mounts up from the atoms and gradually emerges to the level of

our senses, so that those bodies are in motion that we see in sunbeams, moved by blows that remain invisible.'[13]

Particles of flour distributed in air demonstrate Brownian motion.

Nobody took much notice of this idea, and the Dutchman Jan Ingenhousz was puzzled when, in 1785, he noticed particles of coal dust 'dancing' on the surface of alcohol. But he did not follow up the observation, and it was left for the English botanist Robert Brown to carry out the key experiments on what then became known as Brownian motion.

Brown was studying pollen grains through a microscope in 1827. He noticed that when tiny particles (organelles) ejected from these grains were floating in water they moved in a jerky, zig-zag fashion. At first, he thought that these particles might be alive, and swimming about in the water. But as a careful scientist Brown tested this ideas using equally small particles of non-living material, such as cement, and found that small particles of anything suspended in liquid (or, indeed, smoke particles suspended in the air) moved in the same way.

From the 1860s onward, as the idea of atoms gained strength among scientists, there were suggestions that the particles move because they are being hit by atoms or molecules of the liquid in which they are suspended. But in order for a single atom to impart a big enough kick to explain the motion, the atoms would have to be nearly as big as the particles, and should be visible under

Albert Einstein (1879–1955).

the microscope. But later in the nineteenth century, Louis-Georges Gouy in France, and William Ramsay in England, each suggested a statistical mechanism for the phenomenon.

They pointed out that if a particle in a liquid is being buffeted from all sides by tiny atoms or molecules, although on average the force it experiences is the same from all sides, there will be moments when, just by chance, there are more atoms hitting one side of the particle than any other side. So it will jerk away in the opposite direction. Then, another statistical fluctuation will give it a jerk in another random direction. Neither of them did any calculations of how the effect might work, and it was left for Albert Einstein, in 1905, to put the numbers in and provide a precise mathematical explanation of how Brownian motion works. (Einstein hadn't actually read the papers of Gouy or Ramsay, but worked it all out for himself.) He calculated that this zig-zag motion, now known as a 'random walk', would move a particle in random directions with each kick, but that over longer times the distance it moves from its starting point, measured in a straight line, is proportional to the square root of the time that has elapsed since the first kick. So it moves twice as far in four minutes as it does in one minute, four times as far in sixteen minutes, and so on. This prediction was tested in experiments by Jean Baptiste Perrin in 1908, and found to be correct. Einstein wrote to thank him, saying: 'I would have considered it impossible to investigate Brownian motion so precisely; it is a stroke of luck for this subject that you have taken it up.' In 1926, Perrin received the Nobel Prize for his work.

Towards the end of the twentieth century, some people claimed that Brown could not have seen what he claimed to have seen, because his microscopes simply were not good enough. But the microscopist Brian J. Ford then used one of Brown's original microscopes to repeat the experiments from the 1820s, proving that Brown really did observe Brownian motion.

Nº. 33   THE MAGNETISM OF ELECTRICITY

lectricity continued to advance the development of physics, as well as chemistry, in the early decades of the nineteenth century. The Dane Hans Christian Ørsted had been interested in both electricity and magnetism for some time when, on 21 April 1820, he observed something that had not previously been noticed. According to his own account, he was giving a lecture to demonstrate various magnetic and electric phenomena, and had the apparatus for the demonstrations laid out on the table in front of him. He spotted that when a wire was connected to a battery, an ordinary compass needle near the wire flicked over. He had found a direct link between the two phenomena. But he did not draw the attention of his audience to what he had noticed, preferring to carry out some proper experiments before going public.

By moving a magnetic compass needle around to different positions near a wire connected to a battery, Ørsted showed later that year that the electricity in the wire produced a circular magnetic field around the wire. Ørsted was puzzled by what he found. The compass needle was not directly attracted to point to the wire, but neither was it directly repelled to point away from the wire. It seemed to be trying to point perpendicular to it. It pointed in one direction, at right angles to the wire, when the needle was above the wire, and in the opposite direction when the needle was below the wire. In addition, reversing the connections of the wire to the battery reversed the effect. Unable to explain this, Ørsted published a paper describing what he had found and inviting other scientists to come up with an explanation. The challenge was quickly picked up and developed further, not just experimentally but with the first real, albeit incomplete, theory of electromagnetism (as it became known) by the Frenchman André-Marie Ampère.

Ampère learned about Ørsted's discovery in September 1820, when it was demonstrated in Paris by François Jean Arago. This prompted him to carry out his own experiments, in which he showed, among other things, that two parallel wires each carrying an electric current will repel one another if the currents are flowing in opposite directions, but they attract one another when the currents are flowing in the same direction. He found that the strength of this effect is

A wire carrying an electric current generates a magnetic field around it. For a long straight wire the magnetic field lines form circles around the wire, shown here by placing small magnetic compasses around such a wire.

proportional to the strength of the current flowing in the wires and to the length of the wire, and that it obeys an inverse square law – that is, other things being equal, the force is reduced by a factor of four (to a quarter of its original strength) when the wires are moved twice as far apart, and so on. As Ampère himself was fond of pointing out, this is reminiscent of Isaac Newton's 'discovery' of the inverse square law of gravity. You might wonder how he knew the strength of the electric current. It was because he invented an instrument to measure it. The instrument used a freely moving compass needle to measure the flow of electricity by the amount of deflection produced. Later developments of this became known as galvanometers, in honour of Luigi Galvani, while the unit of electric current, of course, was named after Ampère.

Hans Christian Ørsted (1777–1851), seen here with an assistant observing an experiment to demonstrate the effect of an electric current on a magnetic compass needle.

What would prove to be one of the most significant discoveries concerning electromagnetism, as far as its practical implications were concerned, was also made by Ampère, although other researchers independently noticed the same thing. When an electric current is passed through a wire which has been coiled into a helix (dubbed a solenoid by Ampère), it produces a magnetic field that is exactly like the field of a bar magnet, with a north pole at one end of the solenoid and a south pole at the other end. This is the basis of electromagnetic switches, which operate by pulling a strip of metal towards the coil when the current flows, and releasing it when the current is turned off. And it also became the basis of the electric motor (see page 99).

In attempting to provide a unified description of magnetism and electricity, in which he was only partly successful, Ampère introduced the idea of an 'electrodynamic molecule', which carried electric current along a wire; this was the electron in all but name. In 1827 he pulled all his ideas and experimental evidence together in a great book, *Mémoire sur la Théorie Mathématique des Phénomènes Électrodynamiques Uniquement Déduite de l'Experience* (*Memoir on the Mathematical Theory of Electrodynamic Phenomena, Uniquely Deduced from Experience*), in which he introduced the term electrodynamics. James Clerk Maxwell, the British scientist who, later in the nineteenth century, did develop a complete mathematical theory of electromagnetism, wrote (in 1873) that 'the experimental investigation by which Ampère established the law of the mechanical action between electric currents is one of the most brilliant achievements in science'.

Nº. 34   THE DEATH OF VITALISM

The idea of a special life force that operated in the chemistry of living things but not in the chemistry of non-living things took a long time to die. But a key step down that road was taken accidentally in 1828, when the German chemist Friedrich Wöhler found that it is possible to make at least some 'organic' compounds – which had previously been assumed to be produced only by living things – in the proverbial test tube.

Back in 1773, the French scientist Hilaire Rouelle had isolated crystals of a substance that became known as urea from the urine of various animals, including people. This was already something of a puzzle for proponents of vitalism, as urea turned out to be a relatively simple compound (in modern notation, it is $H_2N\text{-}CO\text{-}NH_2$), which hardly seemed complicated enough to require the influence of a life force in its manufacture. Nevertheless, Wöhler was, like most of his contemporaries, a confirmed vitalist when he set out on a course of experiments in which he was trying to make ammonium cyanate by

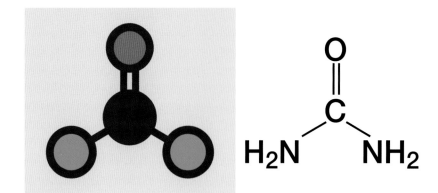

reacting ammonium chloride with silver cyanate. To his surprise, the crystals produced by the reaction were the same as the ones Rouelle had found – urea. (A molecule of ammonium cyanate does actually contain the same atoms as a molecule of urea, but in a different geometrical arrangement.) Wöhler wrote to Berzelius that he could 'make urea without the use of kidneys, either man or dog'. But he was not happy with the discovery, and told Berzelius that 'the great tragedy of science [is] the slaying of a beautiful hypothesis by an ugly fact'. Yet again, if it disagrees with experiment then it is wrong.

This was not, however, the end of vitalism. Urea is a relatively simple 'organic' substance, so perhaps it might be a special case. Many other, more complex compounds still could not be manufactured except by living organisms, and there was widespread resistance to the idea that there was no need for a life force until well into the 1840s. In 1845, however, another German, Adolph Kolbe, deliberately set out to establish that organic compounds could be made from inorganic substances by converting carbon disulphide into acetic acid. As the carbon disulphide itself is easily made out of its constituent parts, the whole process involved a series of chemical steps known as a 'total synthesis'. This is the complete synthesis of an organic compound from simple precursors, without involving any biological processes.

The great champion of total synthesis in the mid-nineteenth century was the Parisian Marcellin Berthelot. In 1855, he synthesized ethyl alcohol from ethylene. Previously, ethyl alcohol had been prepared by fermenting sugars with yeast (a biological process), but ethylene is an inorganic molecule. Berthelot convinced himself that all organic compounds could be manufactured from inorganic substances by a careful process of total synthesis. So he set out on an ambitious programme to synthesize every known organic substance by starting out from simple inorganic substances and building up more and more complicated compounds step by step. He planned to move up from simple hydrocarbons (things like methane) to alcohols (which contain an OH group), esters (in which an OH group is replaced by a more complex 'alkoxy' group), organic acids (which contain even more complex groups), and so on. His successes included the

**Friedrich Wöhler
(1800 –1882).**

synthesis of formic acid (the chemical that ants use to sting with), acetylene (by passing an electric arc like the one invented by Humphry Davy between carbon electrodes in an atmosphere of hydrogen), and benzene (by heating acetylene in a glass tube). The synthesis of benzene was particularly significant because each molecule contains six carbon atoms joined in a ring (see page 132). It occurs naturally in crude oil (the remains of living organisms), but by synthesizing it Berthelot opened up a new branch of chemistry involving the reactions of such ring molecules, now known as aromatics.

Unlike Wöhler, Berthelot was an almost evangelical proponent of the idea that all chemical processes are based on the action of physical forces which can be studied and measured, like the forces involved in mechanical processes. His grand programme of total synthesis was far too ambitious for one man to complete, but he did enough to show that it is indeed possible to create organic substances from the four elements found in all living things – carbon, hydrogen, oxygen and nitrogen, collectively referred to as CHON. When his definitive work on the synthesis of organic chemicals, Chimie Organique Fondée sur la Synthèse (Organic Chemistry Founded on Synthesis), was published in 1860, its sounded the death knell of vitalism.

MAKING ELECTRICITY

T wo key inventions drove the technological revolution. One was the steam engine (see page 48). The other, arguably even more important, was the electric generator/motor combination. Several people contributed to this development, but the key experiments were carried out by Michael Faraday, working at the Royal Institution in London.

Faraday's interest in electricity was stimulated by news of Ørsted's discovery of the magnetic effect associated with an electric current. He realized that a wire carrying an electric current should be forced to move in a circle round a fixed magnet, and in 1821 devised an experiment to demonstrate this effect. A magnet was stuck vertically up through a pool of mercury in a glass container, and a wire was dangled from a support above the dish with its end in the mercury. Mercury is both a liquid and a conductor of electricity, so the wire could move about in the mercury when a current flowed without breaking the electrical circuit. When the wire was attached to a battery via the mercury, the end of the wire in the dish did indeed move in a circle round the magnet, exactly as Faraday had expected. In a variation on the theme, the wire was fixed vertically, but the magnet was free to rotate, and circled round the wire when the current was switched on. This was the genesis of the idea of the electric motor, converting electricity into mechanical motion. But Faraday was diverted into other work, chiefly in chemistry, and in 1825 became Director of the Laboratory at the Royal Institution, where he introduced a series of popular lectures about science. All of this gave him little time for electrical work, and he only returned to the puzzle at the beginning of the 1830s. The question he wanted to answer was: if an electric current can induce a magnetic force in its vicinity, can a magnet induce an electric current in a nearby wire?

By then, several people had found that if a magnetic compass needle is suspended by a thread above a horizontal spinning metal disc the needle is deflected, but the phenomenon was not understood. Faraday's work gave the answer, but what he found was not what he started looking for.

Reconstruction of part of the laboratory belonging to British physicist Michael Faraday (1791–1867), now part of the museum at the Royal Institution, London, in the basement where he worked.

In 1831, Faraday was using an adapted form of a solenoid (see page 101) in his experiments. He had a metal ring made of iron about 2 centimetres thick bent in a circle about 15 centimetres in diameter. On opposite sides of the ring, coils of wire were wound as in a conventional solenoid. One coil was connected to a battery, the other to a sensitive galvanometer. What Faraday was looking for was any deflection of the galvanometer, indicating a flow of current in the second coil, when an electric current was flowing through the first coil. He anticipated that the current in the first coil would induce magnetism in the iron ring, and that the magnetism would induce a current in the second coil.

What he actually found was that when a steady electric current was flowing in the first coil, nothing at all happened to the galvanometer. But on 29 August 1831, to his surprise, Faraday noticed that the galvanometer needle flickered just at the moment the current to the first coil was turned on, then fell back to zero. It flickered again when the current was turned off. Through this and further experiments Faraday established that a steady electric current, producing a steady magnetic field in the ring, did not produce any induced current. But as long as the first current was changing, which means as long as the magnetic field in the ring was changing, a current was induced. He soon found that simply moving an ordinary bar magnet in and out of a solenoid produced a current in the coil as long as the magnet was moving. Just as moving electricity induces magnetism, so moving magnetism induces electricity.

In the spinning disc experiment, what is happening is that because the disc is moving relative to the magnet of the compass needle, electric currents flow in the disc; these currents induce a magnetic field that affects the needle. Faraday's work opened the way to generate electricity more or less on demand by machines that move magnets in circles around suitably coiled wires, or by moving those wires past magnets, since what matters is the relative motion. Just as electricity could be converted into mechanical motion, mechanical motion could be converted into electricity. And electricity could be generated in one place, carried by wires to another place, and used to drive motors there. In May 1881 – just half a century after Faraday noticed the flickering galvanometer needle in his lab – the first electric tram in the world was being tested at Lichterfelde near Berlin.

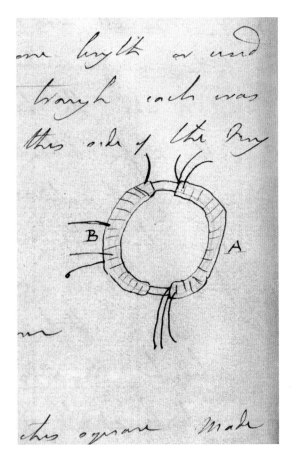

A page from the notebooks of Michael Faraday, dated 29 August 1831, showing notes and a diagram of an electromagnetic induction ring.

AN UPLIFTING EXPERIENCE

Charles Darwin is best known for his theory of natural selection, which explains how evolution works. But before he turned his attention to evolution he was a geologist, and he made his name in scientific circles with his accounts of the geological observations he made in South America during the famous voyage of the *Beagle*. The most dramatic of these concerned his first-hand experience of a major earthquake, the observations of which led him to the correct explanation of the origin of the Andes and other mountain ranges.

Darwin was ashore in Chile, near the town of Valdivia, when he experienced the earthquake, on 20 February 1835. He wrote: '[I] was lying down in the wood to rest myself. It came on suddenly, and lasted two minutes, but the time appeared much longer. The rocking of the ground was very sensible … It was something like the movement … felt by a person skating over thin ice, which bends under the weight of his body. A bad earthquake at once destroys our oldest associations: the earth, the very emblem of solidity, has moved beneath our feet like a thin crust over a fluid.'[14]

Robert FitzRoy, the Captain of the *Beagle*, put his ship out to sea as soon as possible, and headed north along the coast to the city of Concepcion to offer what help he could. They arrived at the devastated city on 4 March. It was near here that Darwin saw along the coastline rock layers raised above the level of high tide, but covered with the remains of dead and dying shell fish, including mussels and limpets, and seaweed anchored to the rocks. It was clear that the earthquake had lifted the land. FitzRoy thought that this must be a temporary phenomenon, and that the land would soon settle back down. But Darwin realized the long-term implications. He had read the work of the

Charles Darwin (1809–1882).

pioneering geologist Charles Lyell, who, in his great book, *Principles of Geology*, made the case that instead of the kind of catastrophes described in the Bible, all that was needed to explain the appearance of the Earth was the accumulation of gradual changes, exactly the same as the processes we see going on around us now, over immense periods of time. He developed these ideas following the work of the Scot James Hutton in the 1780s, according to whom the origins of the Earth were lost in the mists of time, and as far as his observations of the geological record were concerned he could find 'no vestige of a beginning, no prospect of an end'. This flew in the face of the orthodox view at the time, that the Earth was only some six thousand years old. This relatively new idea, known as uniformitarianism, contrasted with the older-established 'catastrophist' view, which held that sudden, dramatic changes unlike anything seen on

Earth today were responsible for the formation of mountain ranges, ocean beds, and other features of the Earth's surface. Lyell was the leading light of uniformitarianism.

Compared with the Biblical Flood, even the earthquake Darwin had just experienced was a gradual change. And it was normal for Chile. If one earthquake could raise the land noticeably, then over the kind of timescale Lyell talked about the whole Andes mountain range could have been lifted out of the sea. The clinching evidence came when Darwin noticed old shell beds higher up the rocks rising from the sea – proof that there had indeed been similar earthquakes producing similar amounts of uplift in the past. On an expedition inland, he even found fossilized sea shells high up in the mountains. All of this would in due course provide Darwin with the time frame needed for evolution to do its work by a similar accumulation of tiny changes over immense intervals. He concluded that 'it is hardly possible to doubt that this great elevation has been effected by successive small uprisings … by an insensibly slow rise'. And when he came to develop his theory of natural selection, Darwin thanked Lyell in print for giving him 'the gift of time' – time enough for natural selection to operate as the mechanism of evolution.

The 'skating on thin ice' analogy also led Darwin to think about what lay beneath the seemingly solid surface of the Earth, and to come to a remarkably modern conclusion: 'In all probability, a subterranean lake of lava is here

Raised beaches, Patagonia, Argentina. From the first illustrated edition of *Voyage of the* Beagle: *Charles Darwin's Journal of Researches* (1890).

stretched out … We may confidently come to the conclusion that the forces which slowly and by little starts uplift continents, and those which at successive periods pour forth volcanic matter from open orifices, are identical. From many reasons, I believe that the frequent quakings of the earth on this line of coast are caused by the rending of the strata, necessarily consequent on the tension of the land when upraised, and their injection by fluidified rock.'[15]

Nº. 37 BLOOD HEAT

One of the most important laws of physics is the law of conservation of energy, which tells us that energy can neither be created nor destroyed, but can be converted from one form into another (in other words, you can't get something for nothing). An everyday example of this happens when a car brakes. The energy of motion of the car (kinetic energy) is converted into heat energy by friction in the brakes. This energy slowly dissipates out into the world as the brakes cool down, but is never lost. The first person to appreciate the significance of the conservation of energy was a German physician, Julius Robert von Mayer, who was the ship's doctor on a Dutch vessel in the East Indies at the beginning of the 1840s.

Mayer's 'experiments' involved cutting the veins of sailors to remove some of their blood. This was routine at the time as a treatment for various illnesses, but it was also a standard practice in the tropics because doctors believed that draining off a little blood would help people to cope with the heat. It was important to open a vein for this blood-letting, rather than an artery, because arterial blood is under higher pressure than veinous blood, and it is harder to control the bleeding from an artery. It is arterial blood, loaded with oxygen from the lungs, that is red; veinous blood, on its way back to the lungs, is a dark purple colour, although it quickly becomes red when exposed to oxygen in the air. When Mayer opened the vein of a sailor in Java, he was astonished to find that the blood was as bright red as arterial blood – indeed, at first he thought he had made a mistake and nicked an artery. But he then tested the rest of the crew, taking great care to cut only veins, and found the same thing.

This highlights the importance of an enquiring mind in science – we might say, a scientific mind. Dozens of doctors must have noticed the bright colour of veinous blood in the tropics before Mayer came along, but he was the first one to see it as a phenomenon worth studying and to carry out experiments to find out what was going on.

Julius Robert von Mayer (1814–1878).

Mayer knew about Lavoisier's work (see page 58), which had established that warm-blooded animals are kept warm by a form of slow combustion, with oxygen from the air combining with components of their food, just as a fire is kept hot by oxygen from the air combining with wood or coal. He correctly inferred that the veinous blood in the tropics was bright red because less oxygen had been used up, as the body does not need to 'burn' so much fuel to maintain its temperature when the outside world is warm. And he made the great leap of understanding to realize that this implied that all forms of energy are interchangeable – heat from the Sun, muscle energy, burning coal, or whatever. Heat, or energy, could never be created but only converted from one form into another.

When he got back to Europe, Mayer wrote up his ideas and tried to publish scientific papers in the hope of drawing attention to them. But unfortunately, because he had no training in physics the papers were difficult to understand and contained errors. At first, he had trouble getting anything published. Even though he then studied physics and published work on the effect of heat on expanding gases, very little notice of his work was taken at the time. It was left for the English physicist James Joule to independently discover the relationship between work and heat in a series of experiments, most famously involving a falling weight connected by a rope and pulley to a paddle-wheel in an insulated barrel of water. The weight falling under gravity converts gravitational energy into the rotational energy of the paddle wheel, which in turn increases the temperature of the water. Mayer's compatriot Hermann von Helmholtz learned about these ideas though Joule's work, and published the definitive statement of the law of conservation of energy in 1847. This was the moment when the idea gained widespread attention, and although Helmholtz had learned about this from reading Joule's publications, eventually Mayer also received the credit that was due to him.

The law of conservation of energy is so important that it is now known as the First Law of Thermodynamics. It means that the total energy of an isolated system does not change, but it is important to remember that the planet we live on is not an isolated system in thermodynamic terms, because it receives energy from the Sun. Mayer, however, was one of the people who

Apparatus to demonstrate how muscle energy can be converted into heat by stirring a bucket of water.

pointed out around this time that according to the best understanding of mid-nineteenth century physics, the Sun itself would run out of energy in a few thousand years. That puzzle would be resolved by the discovery of radioactivity (see page 154).

(see page 154)

Nº. 38   TRUMPETERS ON A TRAIN

Sometimes experiments lead to new discoveries, as with Galvani's twitching frog's legs (see page 61), and sometimes new ideas, or hypotheses, lead to experiments which prove the hypotheses correct and turn them into theories. Often, there is a synergy between the two processes as scientific knowledge advances. That is what happened with one of the two most important tools used by astronomers, now known as the Doppler effect.

Christian Doppler was working at the Prague Polytechnic, the forerunner of the Czech Technical University, in the 1840s, and knew about the relatively recent work of Thomas Young and Augustin Fresnel (see page 78) establishing the wave nature of light. He realized that if a source of light is moving towards an observer, the waves seen by the observer would be squashed to shorter wavelengths (towards the blue end of the spectrum); if it is moving away, the waves would be stretched to longer wavelengths (towards the red end of the spectrum). He thought that this would affect the observed colours of stars, and published a paper drawing attention to this in 1842 (the title translates as *On the Coloured Light of the Binary Stars and some other Stars of the Heavens*). He wrote: 'nothing seems to be more intelligible than that, to an observer, the path length and the interim durations between two consecutive breakings of a wave must become shorter if the observer is hurrying toward the oncoming wave, and longer if he is fleeing from it.'

He was right, but the idea that this would affect the colour of stars turned out to be wrong. Stars do not move fast enough for their colour to be noticeably affected by their motion, although more than a century later a related effect turned out to be important in studies of much more distant objects known as quasars. But the Doppler effect, as he realized, also applies to sound waves moving through the air. If the source of the sound is moving towards you, the sound waves are squashed to a higher pitch; if it is moving away, they are stretched to a deeper note. This effect is very familiar today, as we all notice it when the note of the siren on an emergency vehicle shifts downward (a 'down Doppler') as the vehicle rushes past. But how could the idea be tested in the 1840s?

In 1845, the Dutchman Christoph Buys Ballot devised a brilliantly simple experiment to measure the effect. The fastest vehicle available at the time was a steam train, and he arranged for a group of horn players to play a particular

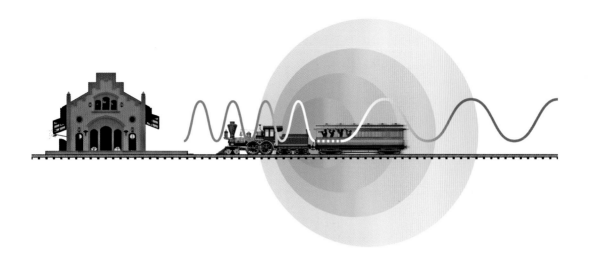

note on the open carriage of a train as it steamed along the Utrecht-Amsterdam line past another group of musicians. These listening musicians had perfect pitch, and were able to describe precisely the way the note changed as the train steamed past them. So the experiment confirmed Doppler's principal hypothesis; but Buys Ballot was also one of the first to point out that the idea that this would affect the perceived colour of stars was wrong.

This, however, was not the end of the story. In 1848, the Frenchman Hippolyte Fizeau realized that certain dark lines in the spectrum investigated by Joseph Fraunhofer would be shifted by the Doppler effect. The origin of these lines was not understood at the time (see page 117), but it was known that they occurred at precise wavelengths in the spectrum of light from the Sun. If the lines were shifted towards the red (redshift), the amount by which they were shifted could be used to measure how fast a star is receding from us, and if they were shifted towards the blue (blueshift), they could be used to measure how fast a star is approaching. This was put into practice by the British astronomer William Higgins in 1868, when he carried out the first measurements of the speed of several stars relative to the Earth. In binary systems, as one star orbits around another, the Doppler effect moves the lines to and fro, giving a measure of how fast the star is moving. Combined with other observations, this makes it possible to measure the masses of stars in binaries – a far cry from listening to trumpeters on a train.

It is worth pointing out, though, that the famous cosmological redshift is not a Doppler effect, even though many popular accounts (and even some textbooks) incorrectly refer to it that way. Although it is measured in the same way, in terms of features in the spectrum of light from distant objects being shifted towards the red end of the spectrum (to longer wavelengths), it is not caused by the motion of those objects (distant galaxies and quasars) through space, but by space itself stretching and stretching the light waves with it as the Universe expands.

The Doppler Effect at work. Sound waves from trumpeters riding on a train are compressed in front of the train, making the pitch of the note higher, and stretched behind the train, lowering the note.

S ome experiments, like Faraday's attempt to make induced electric current with his iron ring, produce results you do not expect, but provide powerful new insights. Others produce exactly the results the experimenter expected, providing powerful evidence to persuade doubters that a new theory is correct. Just such an experiment was carried out in the 1830s by the Swiss geologist Louis Agassiz, to support his contention that the Earth had once been covered by ice.

At that time, there was a debate about the origin of boulders found at various locations across Europe far away from the rock strata to which they belonged. The traditional explanation was that these rocks, known as erratics, had been carried there by the Biblical Flood. But a minority view was that they had been carried to their present locations by ice sheets spreading out from the north and down from the mountains during a great glaciation. Agassiz started out firmly convinced that the Flood story was right, and in 1836 made a field trip to the mountains with the intention of finding evidence to prove the glaciation idea was false. Instead, he soon convinced himself that the distribution of these erratics, the smoothness of rock surfaces polished by the action of glaciers, scratches in the rock scoured by stones dragged along by glaciers, and other evidence supported the idea. He started promoting it, and in 1837 coined the term Ice Age (in German, *Eizeit*). That year, in his capacity as President of the Swiss Society of Natural Sciences, he introduced the term in his Presidential Address, which he used to present the evidence and tell his colleagues that: 'In considering the intimate connection between the different facts which we have just been describing, it is manifest that every explication which does not account at the same time for the polish of the surface of the soil, for the superposition and the rounded form of the pebbles, for the sand reposing immediately upon the polished surfaces, and also for the angular form of the great superficial blocks, is an explication which is quite inadmissible as accounting for the erratic blocks of the Jura; and these objections forcibly apply to all the [other] hypotheses respecting the transport of blocks with which I am acquainted.'[16]

But his colleagues in the Swiss Society of Natural Sciences ridiculed the idea. So he devised an experiment to test the power of ice in moving boulders.

The experiment was disarmingly simple. Agassiz set up a shelter grandly referred to as an observing station (actually a little hut) on a rocky outcrop in the Aar glacier in the Swiss

Louis Agassiz (1807–1873).

mountains. He hammered stakes into the ice nearby, and returned each summer to measure how far the stakes had been carried by the glacier, flowing downhill like a river of ice, and to add more stakes. Over the next three years, he found that the ice moved even faster than he had anticipated, and he observed large boulders being carried along with the ice. He inferred that in the relatively recent past Switzerland had been covered not by a few glaciers on the high Alps, but by a single great sheet of ice, spreading out from the higher Alps and covering the whole valley of northwestern Switzerland. The ice sheet would have been blocked by the Jura mountains, piling up behind them to reach the height of the peaks of the range. And the same sort of thing would be happening across Europe, and beyond – a Great Ice Age.

In 1840 Agassiz published a book spelling out the evidence for an Ice Age and elaborating on the theme. The book, *Étude sur les Glaciers* (*Studies on Glaciers*), was something of a sensation, not least because of Agassiz's dramatic prose:

Shelter built by Louis Agassiz to study the movement of the Aar glacier, Switzerland, 1842.

'The earth was covered by a huge ice sheet which buried the Siberian mammoths, and reached just as far south as did the phenomenon of erratic boulders. This ice sheet filled all the irregularities of the surface of Europe before the uplift of the Alps, the Baltic Sea, all the lakes of Northern Germany and Switzerland. It extended beyond the shorelines of the Mediterranean and of the Atlantic Ocean, and even covered completely North America and Asiatic Russia. When the Alps were uplifted, the ice sheet was pushed upwards like the other rocks, and the debris, broken loose from all the cracks generated by the uplift, fell over its surface and, without becoming rounded (since they underwent no friction), moved down the slope of the ice sheet.'[17]

Also in 1840, Agassiz visited Britain, where he met some of the leading geologists of the day and went on a field trip around Scotland to show them first hand the evidence in support of the Ice Age theory. The two most senior geologists, William Buckland and Charles Lyell, were convinced, and it was arranged that the two of them, plus Agassiz, would present papers on the Ice Age theory at two meetings of the Geological Society of London later that year. These presentations, on 18 November and 2 December 1840, mark the time when the Ice Age theory came in from the cold, although there was at that time no explanation for why the Earth had been so much colder in the past (see page 252).

N⁰. 40 ABSORBING RADIANT HEAT

At the end of the 1850s, the Irish physicist John Tyndall was working at the Royal Institution, where his career overlapped with the declining years of Michael Faraday. Tyndall became interested in the 'radiant heat' (infrared radiation) discovered by William Herschel (see page 71). Previous scientists, notably the Frenchman Joseph Fourier, had speculated that the atmosphere of the Earth acts like a blanket, trapping heat and keeping the surface of the planet warmer than it would otherwise be. But Tyndall put these ideas on a secure footing by carrying out experiments to measure the capacity of different atmospheric gases to absorb infrared radiation.

John Tyndall (1820–1893).

In his experiments, the gas being investigated was in a long brass tube, sealed at each end with transparent rock crystal, which does not absorb in the infrared. One end of the tube was in contact with a heat source. Infrared radiation that had passed through the gas in the tube emerged from the other end onto a sensitive detector called a thermopile, which converts temperature differences into electricity. The other end of the thermopile was in contact with another standard heat source. The amount of electricity produced by the thermopile depended on the difference in temperature between the two

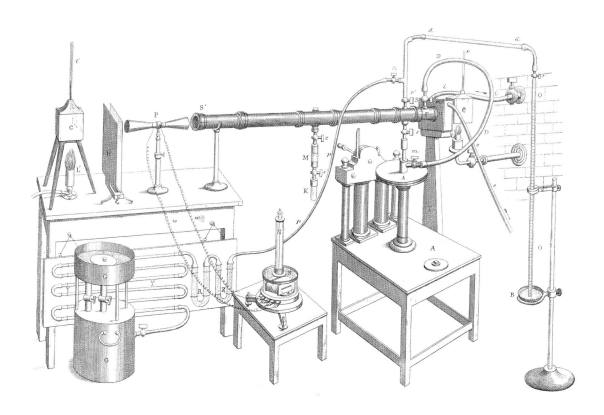

Tyndall's experimental setup to measure the infrared absorbing power of different gases.

ends, and was measured with a galvanometer, to reveal how much (or how little) infrared heat had been absorbed by the gas in the tube. The apparatus, known as a ratiospectrophotometer, was so sensitive that the warmth of a human body would have disturbed the experiment, so the deflection of the needle of the galvanometer had to be read using a telescope from the far side of the laboratory. As Tyndall described it in a lecture at the Royal Institution: 'My assistant stands several feet off. I turn the thermopile towards him. The heat from his face, even at this distance, produces a deflection of 90 degrees. I turn the instrument towards a distant wall, judged to be a little below the average temperature of the room. The needle descends and passes to the other side of zero, declaring by this negative deflection that the pile feels the chill of the wall.'[18]

Tyndall measured the relative infrared absorbing power of nitrogen, oxygen, water vapour, carbon dioxide (then known as carbonic acid), ozone, methane, and other hydrocarbons. He found that the best absorbers of infrared radiation are water vapour, carbon dioxide, and hydrocarbons such as methane. Nitrogen, the main component of the Earth's atmosphere, does not significantly absorb infrared, and nor does the oxygen we breathe, although the tri-atomic form of oxygen, ozone, does. His results were presented to the Royal Society in the Bakerian Lecure of 1861, and later elaborated on in his books and other lectures.

This led Tyndall to develop one of the first scientific explanations for the occurrence of ice ages. Water vapour is the most powerful absorber of infrared radiation, but Tyndall suggested that the amount of water vapour in the air is affected by the amount of carbon dioxide. If there is more carbon dioxide, more heat is trapped and the planet warms a little. That causes more water vapour to be evaporated from the oceans, which enhances the warming. Without the presence of these gases in the air, he said, the Earth would be 'held fast in the iron grip of frost'. This, he thought, could be the explanation for the Ice Age described by Louis Agassiz (see page 108). In the past, there had been less carbon dioxide and less water vapour in the air, and the planet had been colder as a result. Although not the whole story, this is one component of the modern understanding of Ice Age cycles (see page 252), in which small changes in temperature are magnified by such 'feedback' processes.

Tyndall also correctly explained the fall in temperature at night and the formation of dew or frost as a result of heat being lost by radiation, and invented a technique for measuring the amount of carbon dioxide breathed out in each human breath. This technique is still used in hospitals to monitor anaesthetized patients.

The truth of Tyndall's remark about the iron grip of frost is shown by comparing the temperature of the Earth with the temperature of the Moon. The airless Moon receives the same amount of solar light and heat on each square metre of its surface, on average, as the surface of the Earth does. But the average temperature of the Moon (averaged over the entire surface area, day and night) is –18 °C, whereas the similarly averaged temperature of the Earth is +15 °C. Although a little heat does leak out from the Earth's interior, that difference, 33 °C, is almost entirely due to the warming effect of the atmosphere through the trapping of infrared radiation, a process often known today as the greenhouse effect.

Nº. 41 THE LEVIATHAN OF PARSONSTOWN

bservational astronomy began to move in to the modern era in the 1840s, when the third Earl of Rosse (William Parsons) constructed the largest telescope in the world and used it to study objects that are now known to be other galaxies, beyond the Milky Way.

Rosse served as a Member of Parliament from 1822 to 1834, but resigned at the age of 34 to devote himself to astronomy. He was wealthy enough to fund the building of a series of telescopes at the family seat of Birr Castle, in Ireland, culminating in a reflecting telescope with a polished metal mirror (made from 'speculum', a mixture of roughly two-thirds copper and one-third tin) 72 inches

The Leviathan of Parsonstown. Although the telescope was dismantled in 1908, a replica was constructed in the 1990s, and the site is now a museum.

across (6 foot, or 180 centimetres), dubbed the 'Leviathan of Parsonstown'. The barrel of this huge instrument was mounted between two stone towers 15 metres high and 7 metres apart, and could be raised and lowered to point at different heights by a system of ropes and pulleys, but it could not be tracked across the sky and could observe only what swam into its field of view as the Earth rotated. The Earl had to develop the techniques used to build the Leviathan from scratch, both because of its unique size and design and because previous generations of telescope makers had often guarded the secrets of their trade. Unlike them, Rosse was open about his experiments, and presented details of the metal used in making the three-ton mirror – its casting, grinding, and polishing – to the Belfast Natural History Society in 1844, two years after construction of the instrument began. He started observing with the telescope in the following year, 1845, but because of the Irish famine he was preoccupied with his duties as a landowner over the next two years and only began a full programme of observations in 1847.

Rosse was particularly interested in the fuzzy patches of light on the sky known as nebulae (from the Latin for 'cloud'). Using a 36-inch (90 centimetre) reflector, he had previously made drawings (this was before the advent of astronomical photography) of several nebulae, including one which he thought looked a bit like a crab, and named the Crab Nebula. With the 72-inch (180 centimetre) telescope, he was able to see more detail in the nebulae, and in particular he was the first person to see that one of these objects, known as M51, has a spiral structure, like the pattern made by cream being stirred into a cup of coffee. The object became known as he Whirlpool Nebula.

At the time, there were two schools of thought about the nature of these nebulae. One idea was that they were clouds of gas in the process of collapsing under the influence of gravity to make new stars and planets. The other suggestion was that the clouds were made of many stars, too far away and too faint to be picked out individually. Rosse was a firm adherent to the second school of thought, while John Herschel (the astronomer son of William Herschel; see page 56) was a leading light of the first school. It was only in the twentieth century that observations became detailed enough (using even bigger and better telescopes than the Leviathan, and aided by photography) to show that both men were right. Some nebulae are star-forming clouds of gas within our Milky Way Galaxy (some, such as the Crab, are debris from stellar explosions), but others, including the Whirlpool

Whirlpool Galaxy (also known as M51, or NGC 5194), drawn by William Parsons (1800–1867), the third Earl of Rosse, and published in 1850.

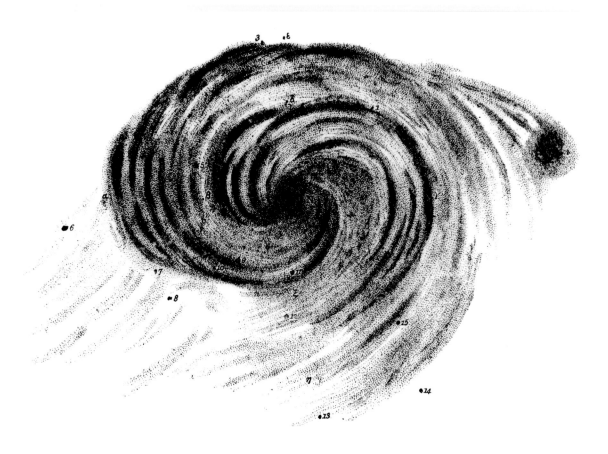

Nebula, are huge star systems similar to, but far beyond, the Milky Way. M51 eventually became the archetypal example of a so-called spiral galaxy.

Of course, the Leviathan could also be used to study objects closer to home, which made a big impression on non-astronomers. The Irish MP Thomas Lefroy said: 'The planet Jupiter, which through an ordinary glass is no larger than a good star, is seen twice as large as the moon appears to the naked eye … But the genius displayed in all the contrivances for wielding this mighty monster even surpasses the design and execution of it. The telescope weighs sixteen tons, and yet Lord Rosse raised it single-handed off its resting place, and two men with ease raised it to any height.'[19]

The third Earl died in 1867, but his son Lawrence, the Fourth Earl, used the telescope until about 1890. Another son, Charles, is famous for inventing the steam turbine. The Leviathan fell into disrepair and was partially dismantled in 1908, but it was rebuilt in the 1990s with a mirror made of aluminium (the original mirror is in the Science Museum in London) and is now a visitor attraction. The first telescope with a larger mirror than the Leviathan, the 100-inch (2.5 metre) Hooker telescope at Mount Wilson in California, did not begin operating until 1918, sixty-three years after Rosse began his pioneering observations.

N⁰. 42   CONTROVERSY AND CONTROLS

Credit in science does not always go where it is deserved, especially when discoveries are 'in the air' and made by different people at nearly the same time. The first use of anaesthesia in surgery is a case in point. Credit is often given to the American William Morton, who applied the technique on 16 October 1846 at the Massachusetts General Hospital, in Boston. He certainly did use the technique, and generated a lot of publicity for it. But he had been pre-empted in 1842 by the less forceful Crawford Long, at his rural medical practice in Jackson County, Georgia. (Crawford Long was a cousin of John Henry 'Doc' Holliday, famous as the dentist friend of Wyatt Earp.) Unlike Morton, Long made a much more thorough and scientific investigation of the anaesthetic technique, including the use of control experiments.

By the early 1840s, it was widely known that inhaling ether could produce sensations of euphoria and even unconsciousness – it was used in some quarters to produce what would now be called a 'legal high', at parties known as 'ether frolics'. It had been used occasionally to relieve the pain of dental extractions, but without any systematic study of what was going on. Long changed all that. On 30 March 1840, in his first operation using ether, the patient was made unconscious using a towel soaked in ether, and Long removed one of two tumours in her neck in front of several medical students.

Replica Morton ether inhaler, similar to the one used at Massachusetts General Hospital in 1846. The glass chamber contained an ether-soaked sponge and the patient inhaled the vapour through the mouthpiece.

She felt nothing, and did not believe he had actually operated until he showed her the tumour. On 6 June, he removed the second tumour.

Over the following months and years, Long carried out further operations, deliberately using ether on some occasions and not on others to confirm that it was indeed the key factor. As he later wrote: 'I was fortunate enough to meet with two cases in which I could satisfactorily test the anesthetic power of ether. From one of these patients, [Mary Vinson] I removed three tumors the same day; the inhalation of ether was used only in the second operation, and was effectual in preventing pain, while the patient suffered severely from the extirpation of the other tumors. In the other case, [Isam] I amputated two fingers of a negro boy; the boy was etherized during one amputation and not the other; he suffered from one operation and was insensible during the other.'[20]

Long was quite open about his work, which was known locally and copied by some of his colleagues, but he did not publish anything or attempt to spread the news until 1849. He even delivered his wife's second baby painlessly with the aid of ether in 1846, the year before James Simpson independently pioneered obstetric anaesthesia in Britain. But William Morton was unaware of any of this when he carried out his well-publicized demonstration of ether anaesthesia in 1846.

Morton had dropped out of both dental school and medical school, but during his brief time as a medical student at Harvard attended chemistry lectures where he learned about the stupefying power of ether. Being unqualified was no bar to practising dentistry in those days, and, on 30 September 1846, Morton carried out a painless tooth extraction using ether. A newspaper report of this success led to an invitation from John Warren, the head surgeon at Massachusetts General Hospital, for Morton to anaesthetise a patient (52-year-old printer, Edward Gilbert Abbott) while Warren operated to remove a tumour from him. This operation, on 16 October 1846, was carried out in front of a large audience of students and surgeons and led to widespread publicity. According to eye-witness accounts, after the operation the patient was asked how he felt and replied 'it feels as if my neck's been scratched'. Within weeks, the news had spread to Europe, where Robert Liston used ether during an operation carried out at University College Hospital in London on 21 December. He commented 'This Yankee dodge beats mesmerism hollow'.

This highlights the second key feature of anaesthesia. Liston was famous as 'the fastest knife in the West End', and could remove a leg in two-and-a-half minutes. Such speed was essential if patients were to survive the agonising ordeal of surgery. But when the patients felt no pain, surgeons could take their time over operations, making fewer mistakes and developing more complex techniques to deal with more complicated problems than removing a tumour from a neck or hacking off a leg.

Nº. 43 FROM FIRE LIGHT TO STAR LIGHT

In 1802 the English scientist William Hyde Wollaston was studying the spectrum of sunlight passed through a glass prism to make a spectrum. He noticed that the coloured patterns were broken up by dark bands, which he thought were simply gaps between the colours, so he did not follow up the discovery. But he did publish the news, and this roused the interest of a German physicist, Joseph von Fraunhofer, who investigated the phenomenon and made a much more detailed study of solar spectra in the second decade of the nineteenth century. Fraunhofer found that instead of a few dark bands, the spectrum was crossed by very many narrow black lines. Eventually, he identified 574 of these separate dark lines. Even more are known today, but all the dark lines in the solar spectrum are known as Fraunhofer Lines. A short section of the spectrum has lines packed together, giving an appearance superficially rather like the lines on a bar code. But what caused them?

The answer, which would ultimately reveal what the Sun and stars are made of, came from the work of Robert Bunsen and Gustav Kirchoff in Germany, in the 1850s and 1860s. This was the same Bunsen whose name is known from the famous burner, and the Bunsen burner was a key feature of the experiments he carried out with Kirchoff.

The fuel for the burner came from the gas that supplied households in Heidelberg at the time. This gas was derived from coal, and it could be combined with oxygen in a Bunsen burner to produce a clear flame. The flame could then be coloured by adding traces of different substances. Chemists used this 'flame test' to identify substances by the colour they give to a flame. Sodium, for example, gives a flame a yellow colour, while copper colours it blue. So when common salt is sprinkled on a flame and turns it yellow, we know sodium is present in the salt. Kirchoff realized that it would be possible to make a more detailed analysis using spectroscopy. So Bunsen

RIGHT **Solar absorption spectrum made by Joseph von Fraunhofer between 1814 and 1815. The black lines (absence of light) are due to the absorption of particular wavelengths of light by chemical elements in the outer layers of the Sun.**

BELOW **Part of the Sun's spectrum, showing Fraunhofer lines. Marked across the top are the alpha, beta, gamma and delta lines of the Balmer series of lines for hydrogen (Wasserstoff), indicating their places in the solar spectrum.**

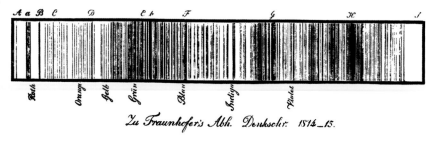

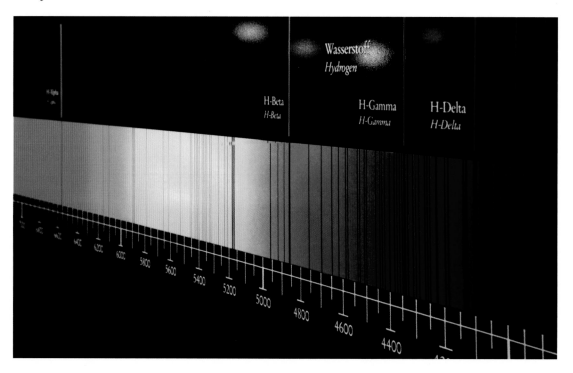

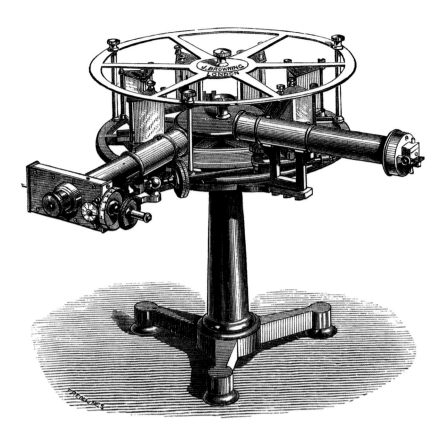

and Kirchoff built an apparatus that included a narrow slit for the light to pass through, a collimator to narrow the beam, a prism to spread the light out into a rainbow pattern, and an eyepiece, like that of a microscope, to view the spectrum. It was the first spectroscope.

When the Heidelbergers analysed the light from flames using spectroscopy they found that each element, when hot, produced bright lines in the spectrum at precise wavelengths – in the yellow part of the spectrum for sodium, in the green/blue part of the spectrum for copper, and so on. One evening, they saw a fire raging in Mannheim, about ten miles away, and from their laboratory in Heidelberg they were able to use the spectroscope to analyse the light from the fire and identify lines produced by the presence of strontium and barium in the blaze.

A few days later, Bunsen and Kirchoff were taking a break from the lab with a walk along the Neckar River, and discussing what they had seen in the fire. Kirchoff later recalled that Bunsen said something along the lines of 'If we can determine the nature of substances burning in Mannheim, we should be able to do the same thing for the Sun.'

'But,' he added, 'people would say we have gone mad to dream of such a thing.'

Following the conversation by the river, they did indeed analyse the spectrum of the Sun, and found that many of the dark lines identified by Fraunhofer were in the same part of the spectrum – at precisely the same wavelengths – as the bright lines produced by various elements when heated in the lab. Kirchoff soon explained what must be going on. These lines were dark because elements present in the outer layer of the Sun are cooler than the layer below. As the light from the hotter layer passes through this cooler region the elements absorb light, removing it from the spectrum at specific wavelengths instead of adding bright lines to it. But the key point is that the lines are in the same places as the bright lines produced when the elements are hot, so the elements can be identified.

Kirchoff's discovery was presented at a meeting of the Prussian Academy of Sciences in Berlin on 27 October 1859. This is now regarded as the date on which the discipline of astrophysics was born, although that name was only coined in 1890. At that time, nobody knew how the lines were produced. But it didn't matter. Even without that understanding, by the 1860s it was possible to find out what the Sun was made of, and the same technique was soon applied to find out what the stars were made of.

Nº 44 PREVENTION IS BETTER THAN CURE

The scientific development of medicine proceeded rapidly in the mid-nineteenth century, with one of the key developments coming from a combination of observation and experiment made by a London-based physician, John Snow. Snow came from the north of England, where in 1831, at the age of 18, he helped out during a cholera outbreak. It was after he moved to London and qualified in medicine that another cholera outbreak, in 1854, led him to an understanding of the way the disease spread.

This was a few years before the work of Louis Pasteur (see page 126) pointed the way to an understanding of germs, and it was widely believed that diseases such as cholera were spread by 'bad air', or miasmas. The cholera outbreak of 1854 centred on the Soho region of London, around Broad Street (since renamed Broadwick Street). It began on 31 August, and within three days 127 people had died. Although most of the local population fled the area, by 10 September 500 people had died, and 616 people died in all.

Snow had doubts about the miasma idea, and decided to identify the source of the outbreak by interviewing local residents and plotting on a map the homes of the victims. This showed that the cholera outbreak was concentrated around Broad Street, where there was a pump from which the residents drew their water. Snow studied the water from the pump using a microscope and analysed it chemically, but could find nothing that might have caused the outbreak. Nevertheless, he was convinced that water from the pump was somehow

John Snow (1813–1858).

Cartoon produced after John Snow linked a cholera outbreak in London to a contaminated pump, showing Death supplying water infected with cholera.

responsible for the cholera, and persuaded the parish authorities to remove the handle from the pump so that it could no longer be used.

As he wrote: 'With regard to the deaths occurring in the locality belonging to the pump, there were 61 instances in which I was informed that the deceased persons used to drink the pump water from Broad Street, either constantly or occasionally … there has been no particular outbreak or prevalence of cholera in this part of London except among the persons who were in the habit of drinking the water of the above-mentioned pump well.

I had an interview with the Board of Guardians of St James's parish, on the evening of the 7th [instance], and represented the above circumstances to them. In consequence of what I said, the handle of the pump was removed on the following day.'[21]

The outbreak declined, although as Snow himself acknowledged it might have been past its peak anyway: 'There is no doubt that the mortality was much diminished, as I said before, by the flight of the population, which commenced soon after the outbreak; but the attacks had so far diminished before the use of the water was stopped, that it is impossible to decide whether the well still contained the cholera poison in an active state, or whether, from some cause, the water had become free from it.'[22]

But support for Snow's hypothesis came from other observations. A nearby workhouse accommodated more than 500 paupers, none of whom suffered from cholera; the workhouse had its own well. And workers at a nearby brewery, who were allowed free beer and never drank water from the pump, were also free from the disease (we now know that fermentation kills the cholera bacillus). Snow later showed that there was an increased incidence of cholera among people who depended on water supplied by the Southwark and Vauxhall Waterworks Company, which was taking water from the sewage-polluted River Thames. But the pump in Broad Street was connected to a well that had been contaminated by sewage from an old cesspit, just under a metre away.

Snow has become known as the founder of epidemiology, but his idea of mapping the cholera outbreak to locate its source may have come from a book by a West-country physician, Thomas Shapter, discussing an outbreak of cholera in Exeter in 1831. Shapter mapped the outbreak essentially without commenting on the possible significance of the map, while Snow used a similar map as a tool.

Snow was also a pioneer in the development of anaesthesia on a proper scientific basis. He was one of the first physicians to calculate proper dosages for the use of ether and chloroform, rather than guessing and hoping for the best, which made the use of anaesthetics in surgery and childbirth safer and more reliable. He personally used chloroform to ease the pain of Queen Victoria when she gave birth to the last two of her nine children, Leopold in 1853 and Beatrice in 1857. She wrote in her diary, 'Doctor Snow gave that blessed chloroform and the effect was soothing, quieting, and delightful beyond measure.' Snow's only comment was *'Her Majesty is a model patient'*.

Nᵒ. 45   PINNING DOWN THE SPEED OF LIGHT

Although the speed of light had been known to be finite since the work of Ole Rømer (see page 40), it was only in the second half of the nineteenth century that the speed was measured with precision. The technique was pioneered by the French physicist Hippolyte Fizeau, and refined by his countryman Léon Foucault, who measured a value within 1 per cent of modern measurements in 1862.

Fizeau made the first accurate ground-based measurement of the speed of light at the end of the 1840s. He used a rotating metal wheel in which gaps had been cut, like the battlements of a castle. A beam of light was fired through a gap in the wheel and along a distance of 8 kilometres between Montmartre and the hilltop of Suresnes, off a mirror and back through the next gap in the toothed wheel. This worked only if the wheel was rotating at just the right speed for the next gap to be in place for the light to get through. By adjusting the speed of the wheel until this worked,

$$\nabla \times E = -\frac{1}{c}\frac{dB}{dt}$$

$$\nabla \times B = \frac{\mu}{c}\left(4\pi i + \frac{dD}{dt}\right)$$

$$\nabla \cdot D = 4\pi\rho$$

$$\nabla \cdot B = 0$$

Maxwell's equations, containing the constant c, which is the speed of light or any other electromagnetic radiation.

Fizeau was able to measure how long it took for light to complete the journey, and thereby get a measure of the speed of light of 313,300 kilometres per second, which is within 5 per cent of the modern determination. He also showed that light travels more slowly through water than through air, which is a key prediction of the wave model. The rival particle model predicts that light travels more rapidly in water.

Foucault also used an experimental method involving bouncing light beams around, but based on a rotating-mirror concept devised by Dominique-François Arago. This technique involved light being bounced from a rotating mirror on to a stationary mirror then back to the rotating mirror. While the light is travelling, the rotating mirror has moved on a little, so the beam of light is deflected by an angle which depends on the distance between the mirrors, the speed of light, and the speed with which the mirror is rotating. By measuring the angle of deflection, Foucault was able to confirm that light travels more slowly in water than in air and then, in 1862, to measure the speed of light as 298,005 kilometres per second. The modern value is 299,792.458 kilometres per second.

This had far-reaching implications. Exactly at the time Foucault was carrying out these experiments, the Scot James Clerk Maxwell was developing a mathematical theory of electromagnetism, building from Faraday's work (see page 99). He found a set of equations that describe all the interactions between electricity and magnetism, and which include a mathematical description of electromagnetic waves. The equations automatically included a constant (now labelled c), which is the speed with which these waves travel through the air. (This was not something that Maxwell put in, it came out of the calculation without

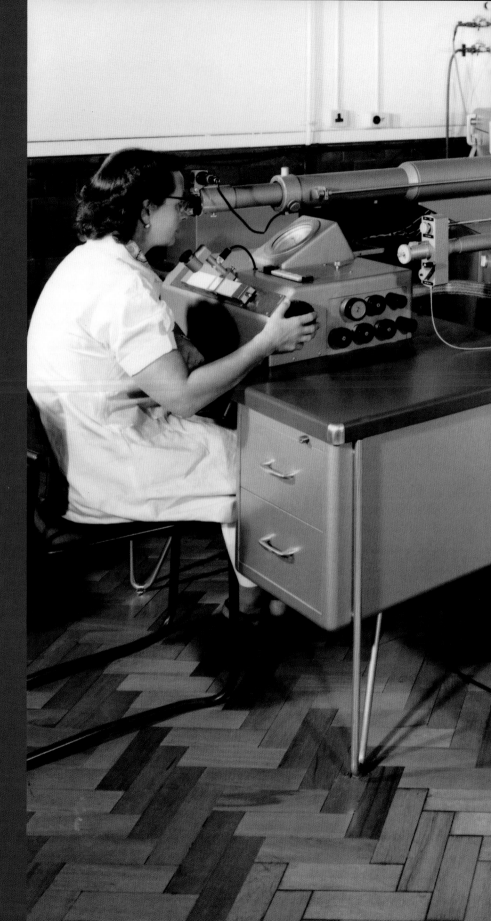

Researchers using an interference comparator to determine the exact length of a metre using microwaves. The metre was redefined in 1983 as the distance travelled by light in a vacuum during 1/299,792,458th of a second. (Photographed at the National Physical Laboratory, Teddington, UK, in 1963.)

being looked for.) That speed was the same as the speed Foucault had measured for light. Maxwell wrote: 'This velocity is so nearly that of light that it seems we have strong reason to conclude that light itself (including radiant heat and other radiations, if any) is an electromagnetic disturbance in the form of waves propagated through the electromagnetic field according to electromagnetic laws.' The 'radiant heat' he referred to is, of course, infrared radiation (see page 110), and the mention of 'other radiation, if any' proved prophetic (see page 143).

However, the equations contained no mention of the speed of the source of the waves, or the speed of the observer measuring them. They seemed to be saying that any observer, moving at any speed, would measure the same speed for any light waves, whatever the source of those waves. This puzzled many people, and strenuous efforts were made to find differences in the speed of light for beams moving in different directions relative to the moving Earth (see page 140). These efforts were unsuccessful, proving that Maxwell's equations were right. Albert Einstein, jumping off from those equations and ignoring the experiments (which he may not even have known about at the time), took the discovery at face value, and made it the cornerstone of his special theory of relativity, published in 1905.

All this is so important that, in 1983, instead of measuring the speed of light in terms of the number of metres travelled in a second, the metre itself was redefined by the Conférence Générale des Poids et Mesures (General Conference on Weights And Measures) as 'the length of the path travelled by light in vacuum during a time interval of $\frac{1}{299792458}$ of a second', setting the value of the speed of light at 299,792,458 metres per second by definition.

Nᵒ· 46   DEATH TO BACTERIA

P roof that diseases are caused by germs invading the body from outside, and refutation of the idea that complex life could arise by 'spontaneous generation' from non-living things, came from the work of Louis Pasteur in France in the second half of the nineteenth century. In 1854, he became Professor of Chemistry at the University of Lille, where his duties included advising local industries on overcoming practical problems. The work for which he became famous stemmed from his work with the brewing industry, finding ways to prevent products going sour.

Pasteur studied sour beer under the microscope, and found that it was filled with a large number of tiny organisms. It had previously been thought that such contamination was produced by the beer going sour; but Pasteur was convinced that they were the cause of the beer going sour. He showed that fermentation is produced by naturally occurring yeasts, and that grape juice drawn by a

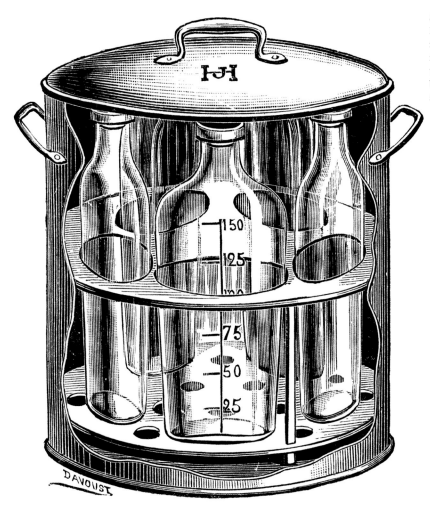

Sterilization of milk. Nineteenth-century cutaway illustration, showing the inside of a rack designed to be filled with milk bottles and sealed prior to sterilization.

hypodermic needle from inside the fruit would never ferment in a sterilized container, because the yeast was present only on the skin of the fruit. The sourness in wine and beer that had 'gone off' was being caused by contamination from outside, which triggered the production of lactic acid in the brew. The same problem occurred with milk. Pasteur found that heating milk to a temperature between 60 °C and 100 °C, and then cooling it, killed the offending organisms (bacteria), preventing it going sour. The process, tested in 1862, became known as Pasteurisation.

But this was just a beginning. Although Pasteur was not the first person to come up with the germ theory of disease, it was a minority view in the 1850s, vociferously opposed by the medical establishment. Pasteur promoted the idea and proposed that disease could be prevented by stopping microorganisms entering the body. This encouraged others to develop the use of antiseptics and promote cleanliness in surgery, leading to a dramatic decline in hospital death rates. In another series of experiments, Pasteur boiled broths in glass flasks to 'Pasteurise' them, then left them connected to the air through narrow tubes with filters to prevent any contamination getting in. Nothing grew in the broths. He wrote: 'Never will the doctrine of spontaneous generation recover from the mortal blow of this simple experiment. There is no known circumstance in which it can be confirmed that microscopic beings came into the world without germs, without parents similar to themselves.'

Pasteur went on to work with diseases, and discovered the technique of preparing a weakened form of a disease-causing agent to act as a vaccine. This was different from Edward Jenner's pioneering work (see page 68), which had used a natural but weaker disease (cowpox) to provide immunity against a deadly disease (smallpox). In 1879 Pasteur discovered that after chickens were injected with a culture of chicken cholera that had accidentally been left to one side for a month they became unwell, but recovered. His assistant was going to throw away the 'faulty' culture, but Pasteur stopped him. He realized that the chickens might now be immune to the disease, and that proved to be the case. Pasteur went on to develop vaccines for anthrax and rabies in a similar fashion, weakening (or killing) the bacteria in various ways, and was responsible for suggesting that Jenner's term 'vaccine' should be applied to all such artificially weakened disease organisms.

The work on rabies led Pasteur into a professional indiscretion that could have had serious consequences. The vaccine was developed by a medical doctor, Emile Roux, who was a colleague of Pasteur. It had been tested successfully on 50 dogs when, on 6 July 1885 a nine-year-old boy, Joseph Meister, was bitten by a rabid dog and brought to Pasteur. After discussing the situation with his team, and with time of the essence, Pasteur personally administered the vaccine to the boy: 'The death of this child appearing to be inevitable, I decided, not without lively and sore anxiety, as may well be believed, to try upon Joseph Meister, the method which I had found constantly successful with dogs. Consequently, sixty hours after the bites, and in the presence of Drs Vulpian and Grancher, young Meister was inoculated under a fold of skin with half a syringeful of the spinal cord of a rabbit, which had died of rabies. It had been preserved (for) fifteen days in a flask of dry air. In the following days, fresh inoculations were made. I thus made thirteen inoculations. On the last days, I inoculated Joseph Meister with the most virulent virus of rabies.'[23]

Pasteur was not a qualified medical doctor, and would have been in serious trouble if things had gone wrong. Fortunately, however, the boy recovered, and the positive publicity helped to ensure the widespread acceptance of vaccination.

Charles Darwin developed his theory of natural selection from observations of the living world. But he also carried out experiments to test that theory. Some of the most important of these were detailed in his book, *On the Various Contrivances by which British and Foreign Orchids are Fertilised by Insects, and on the Good Effects of Intercrossing* (usually known simply as *Fertilisation of Orchids*), published in 1862, three years after the publication of *On the Origin of Species*.

'Star of Bethlehem' orchid (*Angraecum sesquipedale*), also known as Darwin's orchid. This species of orchid was discovered by Darwin during his journey on the HMS Beagle. The nectar tubes of the orchid can measure up to 30 centimetres long. The orchid attracts the hawk moth, *Xanthopan morganii praedicta*, which has a proboscis 30 centimetres long, used to feed on and pollinate the flower.

Fertilisation of Orchids provided the first detailed explanation of how natural selection affected the coevolution of orchids and insects. Although it did not sell to the general public in the same numbers as the *Origin*, it became a classic among biologists and has influenced their thinking all the way to the present day. Darwin's work involved observations, dissection of plants to examine their inner workings, and experiments in which plants were fertilized by hand, with pollen being transferred from one flower to another. His superbly accurate dissections revealed previously unknown features of these plants, including the discovery that members of the genus *Catasetum*, which had very different flowers and had been thought to be entirely separate species, are actually male and female forms of the same plant.

In the 1850s, Darwin gathered evidence to confirm the idea that plants are fertilized by insects transferring pollen from one flower to another as they feed, so the plants do not fertilize themselves. This is an important element of the theory of natural selection, because such cross-pollination provides the variety on which evolution acts. The offspring inherit some characteristics from each parent, giving the potential for them to be slightly different from their parents. Advantageous differences spread and become common; damaging differences die out.

Although he seldom saw insects in action pollinating the orchids near his home in Kent, Darwin was able, by examining the flowers at regular intervals, to discover when pollen had been removed, confirming that they had been visited by insects. In the *Origin*, he introduced the idea of coevolution – whereby insects and plants become adapted to each other as the generations pass – and wrote 'a flower and a bee might slowly become, either simultaneously or one after the other, modified and adapted in the most perfect manner to each other, by the continued preservation of individuals presenting mutual and slightly favourable deviations of structure'. Insects evolve to be better able to extract nectar from the flowers, and the flowers evolve to be more efficient at sticking pollen to the insects as they feed.

Darwin obtained different orchids from correspondents around the British Isles and beyond. This led to a dramatic discovery and a testable scientific prediction. When an insect lands on the large projecting lower petal of an orchid, it pushes its head and tongue (proboscis) down into the centre of the flower to get to the nectar, giving the plant the opportunity to stick pollen masses to the insect. (The proboscis is coiled up, except when the insect is feeding, but flicks out when needed.) If it is too easy for the insect to reach the nectar, the pollen never gets stuck to it. But if it is too difficult, the insect will not come to feed. So plants evolve to make it difficult, but not impossible, for the proboscis to reach the nectar. Variation means that in each generation plants for which this is impossible do not survive, and plants for which it is too easy do not produce as many offspring as plants for which it is just right. At the same time, insects with longer tongues will do better at getting nectar and will thrive, leaving more offspring.

As tongues get longer, evolution of the flowers favours less-accessible nectar; as nectar gets less accessible, evolution of the insects favours longer tongues. So there is a kind of arms race. Plants gradually evolve less-accessible nectar, and insects gradually evolve longer tongues – coevolution. In the end, one kind of plant and one kind of insect become so well adapted to each other that neither can survive on its own. The insect feeds only on one species of plant, and the plant is fertilized only by that species of insect. Darwin's experiments and dissections showed that the nectar of the 'Star of Bethlehem' (*Angraecum sesquipedale*, an orchid from Madagascar), could be reached only down a tube with the 'astonishing' (his word) length of 11½ inches (290 mm). When he first examined samples of the flower, he wrote to a friend with the news, exclaiming 'Good Heavens what insect can suck it'. It implied the need for an as yet unknown moth with a proboscis 10 to 11 inches long to pollinate these flowers. He wrote in *Fertilisation of Orchids*, 'there has been a race in gaining length between the nectary of the *Angraecum* and the proboscis of certain moths'. The insect, a variety of hawk moth, was found in 1903, 21 years after Darwin died. It was named *Xanthopan morgani praedicta*, because it had been predicted by Darwin.

Darwin's sphinx moth, found in Madagascar.

THE BENZENE SNAKE DANCE

The discovery that the molecule benzene is built around a ring of carbon atoms joined to one another as if holding hands opened up a new branch of chemistry that would eventually provide key insights into the nature of the molecules of life. Credit for the image of the benzene ring goes to the German August Kekulé, who published the idea in 1865. But the experimental work on which the discovery was based had largely been carried out by the Scot Archibald Scott Couper, working in Paris in the 1850s, and Joseph Loschmidt, a German based in Vienna, a little later. Although Kekulé did his own experiments, these were similar to those of the other researchers, and, as Couper came first, we will describe his approach here.

The compounds that were important in this work were known as 'aromatics', because of their pleasant smell. They could be derived from benzene and their molecules are, we now know, built around the benzene ring. As a result, all compounds containing a benzene ring, or indeed other flat rings, are now called aromatics, whatever they smell like.

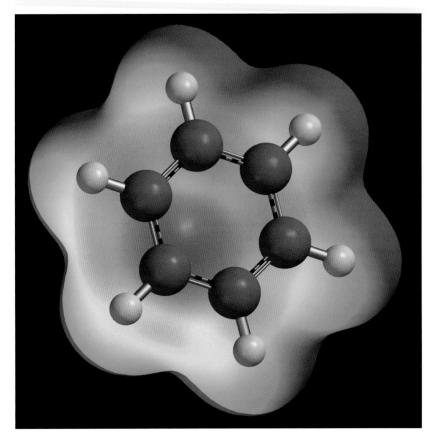

Representation of a benzene molecule. Carbon atoms are shown as black, hydrogen as grey. The bonds between the carbon atoms are shown as 'one and a half', averaging over alternating double and single bonds.

Benzene has a very unusual chemical formula, C_6H_6, meaning each molecule is made up of six carbon atoms and six hydrogen atoms linked in some way. Just how they were linked was a mystery prior to 1865, but as a step towards working out the structure Couper experimented with ways to convert benzene into related compounds containing the hydroxyl group (or radical), OH. These compounds include C_6H_7OH and $C_6H_6(OH)_2$. He also studied salicylic acid, $C_6H_4(OH)COOH$, and tried to find theoretical structures that could explain the properties of these compounds. In doing so, he was the first person to realize that carbon atoms could link to one another to form chains, and the first to attempt to depict the physical arrangement of carbon atoms within the benzene molecule. Early in 1858, Couper wrote a paper on his work and gave it to the head of his laboratory, Charles Adolphe Wurtz, to pass on to the French Academy of Sciences. Because Wurtz sat on the paper for several weeks, it was presented to the Academy only in June, a month after Kekulé published a paper announcing similar results. So Kekulé got the credit that Couper thought he deserved. This led to a furious row with Wurtz, who threw him out of the laboratory; Couper never developed the ideas further, leaving the way open for Kekulé.

Friderich August Kekulé von Stradonitz, usually known as August Kekulé (1829–1896).

By that time, the idea that atoms have a certain 'valence' had been established. This is a measure of the ability of an atom to make chemical bonds with other atoms. Hydrogen has a valence of one, so can form one bond; oxygen has a valence of two, so can form two bonds. In a molecule of water, H_2O, one oxygen atom is linked to two hydrogen atoms, one for each bond, represented as O-H-O. Carbon has a valence of four, so it is easy to understand the structure of methane, CH_4. But a key feature of Couper's work was his willingness to consider the possibility that carbon atoms sometimes behave as if they have a valence of two. This occurs when a double bond links carbon to other atoms, as in carbon dioxide, CO_2, which can be represented as O=C=O. Kekulé was against the idea at first, but it became a key feature of his revelation about the benzene ring.

'Revelation' is the right word, because, according to Kekulé, the solution to the puzzle came to him in a dream. In 1861, Loschmidt published drawings of suggested structures for several molecules, including some aromatic compounds. But in the case of the aromatics, he drew a circle where the benzene itself should be, to represent the unknown structure of the molecule. He did not, as has sometimes wrongly been claimed, suggest that the structure really was circular. It may be, though, that these drawings influenced Kekulé subconsciously. Kekulé published his idea of the benzene molecule as a ring in 1865; much later, he said that the idea came to him in day-dream in which he pictured a snake biting its own tail, like the mythical worm Ouroboros. The idea works because in a ring of six carbon atoms each atom can be joined to the neighbour on one side by a single bond and to the neighbour on the other side by a double bond, leaving the fourth bond free to link up with an atom of hydrogen or with any other atom, or to link to other rings. The importance of this discovery, which opened up a new area of chemical research, is highlighted by the fact that the molecules of life, DNA and RNA, are built around aromatic rings (see page 224).

Nº 49   THE MONK AND THE PEAS

Sometimes the significance of experiments is not widely appreciated at first, either because they receive little publicity, or because they do not fit into the framework of current thinking – or both, as in the case of Gregor Mendel's investigation of inheritance in peas.

Mendel was a monk, based at a monastery in Brno (then part of the Austro-Hungarian Empire but now in the Czech Republic). He was also a trained scientist who had studied at the University of Vienna. This was not an unusual combination in those days – the same monastery also had a botanist and an astronomer among its ranks, and it was an intellectual as well as a religious

Round peas (right) and wrinkled ones (left). Pea traits like these were investigated by Gregor Mendel.

centre. Mendel's main role in the community was as a teacher at the local school, but from 1856 onwards he also had time to carry out a series of experiments on the way in which heredity works in pea plants.

He chose peas because he knew that they had distinctive characteristics that bred true and which could be analysed statistically. The characteristics he studied included whether the seeds were rough or smooth, whether they were yellow or green, and so on. What made this work special, for the time, was that he approached the study of biology like a physicist, carrying out repeatable experiments, keeping detailed records, and applying proper statistical tests to analyse what he saw. He started out with some 28,000 plants, and chose 12,835 for detailed investigation. Each plant was identified as an individual, and records of its descendants were kept like a family tree. He had to know the parents, grandparents, and so on of each plant in succeeding generations, so he had to fertilize every flower of thousands of plants by hand, dusting the pollen from a specific single plant on to the flowers of another specific single plant. Then, he had to analyse the nature of the resulting seeds, plant them, tend the next generation of plants as they grew, and repeat the whole process. It took him seven years to determine the way the characteristics he was studying were passed on from one generation to another.

Austrian botanist Gregor Johann Mendel (1822–1884).

Mendel studied seven characteristics in all, but we can understand what he found using the rough/smooth example already mentioned. He discovered that there is something in a plant that is passed from one generation to the next and determines the nature of the offspring. That something, we now know, is a package of genes, and we shall use the term to describe what Mendel found, even though he did not use it himself (he referred to hereditary elements). Mendel's statistics showed that the properties he studied related to pairs of genes. There is a gene for roughness, R, and a gene for smoothness, S. A single plant must inherit one possibility from each parent, so it may contain any of the possible combinations RR, RS, or SS. It passes one of these on to the next generation. An RR or SS plant has no choice but to pass on R or S, respectively. But an RS plant will pass on R to half its offspring and S to the other half. RR plants always have rough seeds. SS plants always have smooth seeds. But what happens in RS plants? The statistics showed that in this case the R is ignored and the peas are all smooth.

The evidence came from crossing plants that always produce rough seeds (RR) with plants that always produce smooth seeds (SS). Only 25 per cent of the offspring had rough seeds, while 75 per cent had smooth seeds. This is because 25 per cent of the offspring are RR, 25 per cent are SS, and the rest, adding up to 50 per cent, are either RS or SR, both of which give smooth seeds.

Mendel's results were published in 1866, but their significance was not appreciated. It was only at the end of the nineteenth century, when other researchers independently discovered the same laws of inheritance, that his papers were rediscovered and he was given the credit he deserves. The laws of inheritance that Mendel discovered are of key importance in understanding the theory of evolution by natural selection. First, they explain why offspring do not have properties that are a blend of the characteristics of their parents. Offspring of a cross between R and S plants always come out either rough or smooth, not some slightly wrinkled middle compromise. This had been a key puzzle since the publication of Darwin's *Origin* in 1859, because such blending would remove (or at least reduce) the variability on which natural selection operates. Secondly, Mendel showed that each characteristic is inherited independently. Whether or not the pea is green or yellow, for example, does not affect whether it is rough or smooth. The next step towards an understanding of the mechanism of evolution would be taken by Thomas Hunt Morgan, early in the twentieth century (see page 173).

N⁰. 50 THE IMPORTANCE OF NOTHING

One of the most important inventions of the nineteenth century, leading to the discovery that atoms are not indivisible, came about through experiments with the behaviour of electricity in a vacuum. Julius Plücker, a German physicist working in Bonn in the 1850s, wanted to study electricity in this way, and got his colleague Heinrich Geissler, a skilled glassblower, to make suitable glass vessels to use in these experiments, and a pump to suck air out of them. Wires were sealed in to the tube at each end, connected to metal plates, but with a gap between them. When they were connected to a source of electricity, usually an induction coil developed from the work of Michael Faraday (see page 99), electricity flowed from the cathode across the empty space in the tube to the anode. This was like the way electricity flowed through a liquid in Humphry Davy's experiments (see page 85), even though there was only a trace of air in the tube.

Plücker noticed that whatever it was that was coming out of the cathode made the glass of the 'Geissler tube' glow where it struck, near the anode, and that the spot of light could be moved about using a magnet. He also discovered that traces of gas in the tube, such as neon or argon, made the whole inside of the tube glow with different colours, and was the first person to realize that lines in the spectrum of this light were related to the elements producing the light, although he did not develop this discovery as fully as Robert Bunsen and Gustav Kirchoff (see page 117).

By the 1880s, Geissler tubes were being manufactured and sold as ornamental devices for entertainment (a bit like the more recent lava lamps). They came in a variety of shapes, including helical tubes, ones with several spherical bulbs placed along the tube (like a string of onions), and more fancy designs, containing traces of different gases to make pretty colours. Early in the twentieth century, it was realized that this could be adapted for commercial application, and by 1910 coloured tubes were being used in advertising signs. These became generally referred to as neon tubes, or neon signs, even though other gases are also used in them.

Meanwhile, science was developing the Geissler tube for other purposes. By the early 1870s the English physicist William Crookes, based in London, had developed an improved version of the Geissler tube, which became known as the Crookes tube. The key improvement was that Crookes was able to evacuate his tubes to a lower pressure than Geissler, using a pump made by Charles Gimingham. A Geissler tube operated with a pressure of the gas inside of about one-thousandth of atmospheric pressure, but in a Crookes tube the

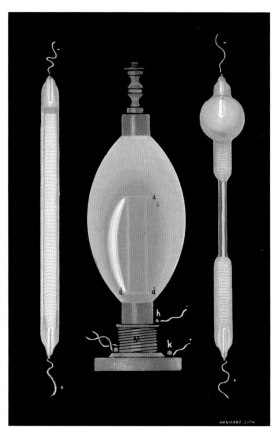

Electric discharges in Geissler tubes.

pressure could go down to a few hundred-millionths of an atmosphere, nearly a hundred thousand times thinner than the gas in a Geissler tube. This made it possible for Crookes and other researchers to probe further into what was going on. As they pumped more air out of his tubes, a dark area, now known as the Crookes dark space, developed near the cathode in the glowing gas. As the pressure went down, the dark area spread down the tube, but the glass behind the anode began to glow. The anode itself left a clear shadow in this glow, with sharp edges.

It was clear that something was being emitted by the cathode, and that it must be travelling in straight lines to leave the shadow. That 'something' became known as cathode rays, a name coined by the German physicist Eugen Goldstein in 1876. As more air was pumped out of the tube, there were fewer gas molecules to get in the way of these rays, so they could travel further before they hit one and made it glow. Eventually (at low enough pressure) they could travel in straight lines from the cathode to the anode, but many flew past the anode and hit the glass of the tube. A particularly neat example of this was demonstrated by Plücker, who made an anode shaped like a Maltese Cross which cast a sharp cross-shaped shadow on the fluorescence behind it.

Modern version of a 'Crookes tube' experiment showing the shadow cast by a metal cross placed in the path of a beam of electrons. The shadow shows that cathode rays travel in straight lines.

Practical applications of developments from these tubes in the twentieth century led to the invention, in 1906, of the vacuum tubes (electronic valves) that preceded transistors in radio, TV and other amplifiers, and later the cathode ray tubes that formed the screens of early televisions. In science, they were instrumental in the discovery of X-rays (see page 150) and the electron (see page 152). Cathode rays are indeed beams of electrons, which are accelerated in a Crookes tube to a very high velocity of about 60,000 kilometres per second, which is roughly a fifth of the speed of light.

N⁰. 51   FEELING THE SQUEEZE

In the middle of the eighteenth century, Carl Linnaeus and other researchers had discovered and studied the pyroelectric effect, which produces an electric potential difference (a voltage) across the opposite sides of certain kinds of crystal when they are heated. It was suggested that the effect could be explained as a result of the warmth stretching the crystals in such a way that the balance between negative and positive charges inside them was distorted.

This led several people to come up with the idea that a similar electrical effect might be produced by squeezing crystals physically; but it was not until the beginning of the 1880s that the French brothers Jacques and Pierre Curie carried out experiments that proved that this was the case, and measured the tiny voltages produced in crystals of quartz, tourmaline, topaz, cane sugar, and Rochelle salt (potassium sodium tartrate). They gave the phenomenon the name piezoelectricity (from the Greek word *piezein*, meaning '*squeeze*').

In the following year, 1881, the Luxembourgian Gabriel Lippmann predicted that the opposite effect should also occur – that if an electric potential is applied across the crystal it will be distorted. Almost immediately, the Curie brothers carried out a further series of experiments which showed that this is indeed the case. As the voltage applied to the crystal varies, the crystal stretches and squeezes in response.

The effect occurs not only in the kinds of crystals (such as quartz) that the word 'crystal' conjures up to most people, but also in other substances, including in some ceramics and even biological material such as bone, which are 'crystals' to scientists because of the repeating pattern of atoms in the structure, like a repeating pattern in wallpaper. The crystal pattern is made up of units (charged atoms, or ions) which each have positive or negative charge, arranged so that the overall charge cancels out. Substances that exhibit piezoelectricity have a particular pattern of repeating atoms, which makes it possible for the squeeze to push some of the atoms closer together or further apart, changing the balance of charges so that one side of the crystal ends up with a positive charge and the other side with a negative charge. This is a very small effect. In a typical crystal, the deformation is about 0.1 per cent of the crystals original size.

The piezoelectric properties of bone have an important biological role. If a bone is under stress this produces a local piezoelectric effect in the stressed region. The resulting electric charge attracts the bone-building cells known as osteoblasts, which deposit calcium and other minerals on the stressed region of the bone, increasing the density of the bone where it is most needed.

Until the First World War, piezoelectricity was little more than a curiosity. But the reverse piezoelectric effect, when a crystal is made to vibrate using electricity, suddenly became important when the French physicist Paul Langevin (one of Pierre

French physicist Paul Langevin (1872–1946), right, who studied under Pierre Curie.

Early 1920s' record player.

Curie's former students) and his colleagues used it to generate ultrasound waves which could be used to detect submarines – the forerunner of modern SONAR devices. At about the same time, unknown to the French team, Ernest Rutherford and his colleagues in Manchester also developed a similar detector for hunting submarines. The vibrating crystal could generate a chirrup of ultrasound which travelled from a ship through the water to bounce off a submarine and be detected back at the ship by a hydrophone. Such an ultrasound generator is called an ultrasound transducer. The development of the device came too late to affect the progress of the First World War, although Langevin's team did detect echoes from a submarine at a depth of 1500 metres in 1918. It was, though, a crucial tool in the battle against the U-boats in the Second World War.

Langevin's work is widely regarded as having kickstarted the study of ultrasonics and its applications. But since 1918, piezoelectricity has also been used in many other applications. In microphones, vibrating sound waves are used to jostle a crystal, generating electricity which can go to an amplifier or recording device. The same sort of thing happens in an old-fashioned record player, where the needle is vibrated as it moves along the grooves in the vinyl record, producing a variable electric current. The opposite effect (also known as a transducer) happens in a quartz watch or clock, where electricity from a battery makes a crystal vibrate very rapidly at a precisely known rate. This is then slowed down to a steady tick electronically, and used to move the hands of the timepiece around its face. And in the most basic application, spark ignition lighters used to fire up gas rings operate by piezoelectricity; when you press the trigger a crystal is squeezed, generating an electric potential difference that makes a spark jump across a narrow gap.

Nᵒ· 52   THE SPEED OF LIGHT IS CONSTANT

After it had been established that light travels as a wave (see page 78), and the speed of light had been measured accurately enough to show that it obeys Maxwell's equations for an electromagnetic wave (see page 123), the natural assumption was that there must be some substance through which the waves travelled, just as sound waves travel through water or air – or, indeed, through solids. This mysterious substance was called 'the ether'. It was assumed that Maxwell's equations were telling us the speed of light through the ether. But this substance would have to have some very curious properties. The speed of a wave through any material depends on the stiffness of the material (sound, for example, travels more quickly through steel than through air). Because the speed of light is so great (just under 300,000 kilometres per second), the ether must be extremely stiff – much stiffer than steel. On the other hand, because planets and other objects move through the ether without being impeded, it had to be very tenuous.

Putting this puzzle to one side, in the 1880s the American physicist Albert Michelson set out to measure the way the Earth is moving through the ether. He did his first experiments while working in Berlin in 1881, but then teamed up with Edward Morley in Ohio to carry out the definitive version of what became known as the Michelson-Morley experiment, completed in 1887.

The idea behind the experiment used interference between beams of light, echoing the way Thomas Young and Augustin Fresnel had established that light is a wave. Using a system of prisms and mirrors, a beam of light could be split in two. The two resulting beams could each be bounced between a set of mirrors and back to a detector, with each beam having travelled exactly the same distance but by a different route. Michelson reasoned that because of the Earth's motion through the ether, the beams would take different times to complete their journeys, and get out of step with one another, so they would interfere to make a pattern of light-and-dark stripes like the one in the double-slit experiment (see page 79). The most extreme effect would occur when the beams were travelling at right angles to each other, with one beam travelling across the direction of the Earth's motion and the other in the same direction as the Earth's motion. The first beam should be unaffected by the relative motion of

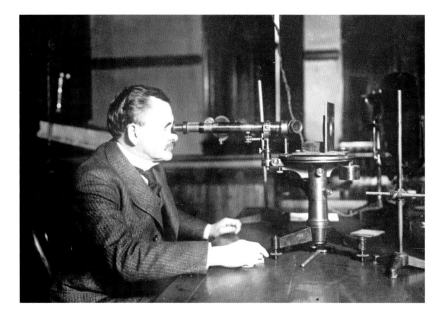

Albert Michelson (1852–1931) looking through a spectrograph.

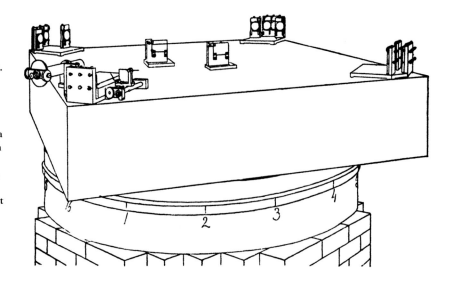

Diagram of the apparatus used by Albert Michelson and Edward Morley in an attempt to detect ether drift. The apparatus consisted of a stone block resting on a pool of mercury contained in a round iron trough. A light source (not seen) sent a beam of white light through a semi-silvered mirror near the centre of the block. The mirror split the light into two beams travelling at right angles to each other. The beams reached the corners of the block and were reflected back to the centre by mirrors. They travelled through the central mirror and out to a telescope lens (not seen), where they formed interference patterns.

the Earth and the ether, while the second beam felt the full influence. But it would require exquisite care in building the apparatus, known as a Michelson interferometer, and carrying out the observations.

The whole apparatus was mounted on a block of sandstone floating in a bath of mercury, with hardly any friction. When the sandstone was given a gentle push, it would slowly rotate in a circle, while the experimenters watched the pattern of interference fringes to see if there was any change. This meant that they could monitor the entire possible range of orientations relative to the ether in a few minutes. They found nothing. In developments from the first experiment, they made observations at different times of day, and at different times of year, to see if there was any observable effect due to the Earth's rotation or its orbital motion around the Sun. They still found nothing.

In a letter to Lord Rayleigh, a British physicist, Michelson wrote: 'The Experiments on the relative motion of the earth and ether have been completed and the result decidedly negative. The expected deviation of the interference fringes from the zero should have been 0.40 of a fringe – the maximum displacement was 0.02 and the average much less than 0.01 – and then not in the right place. As displacement is proportional to squares of the relative velocities it follows that if the ether does slip past the relative velocity is less than one sixth of the earth's velocity.'[24]

The experiment and its 'null result' profoundly influenced science in two ways. First, it was a key step in establishing the idea that the speed of light is (as Maxwell's equations said it must be) an absolute constant, the same for all observers no matter how they are moving. The constancy of the speed of light was one of the postulates on which Albert Einstein would base his special theory of relativity in 1905, although he jumped off directly from Maxwell's equations and was not particularly influenced by the Michelson-Morley experiment. Secondly, the experiment was a key step in the eventual death of the idea of the ether, and its replacement by a field theory in which electromagnetic influences propagate through empty space. In 1907 Michelson received the Nobel Prize for 'his optical precision instruments and the spectroscopic and metrological investigations carried out with their aid'.

N⁰. 53   SPARKING RADIO IN TO LIFE

E lectromagnetism, Maxwell's equations, and light were all cornerstones of scientific experiment and research in the late nineteenth century. At the same time that Michelson and Morley were carrying out their experiments in what proved a futile attempt to find evidence for the ether, the German physicist Heinrich Hertz was making a breakthrough which not only provided further proof of the accuracy of Maxwell's theory, but opened the way to modern communications.

When Maxwell found that the equations that describe electromagnetic waves determined that these waves must travel at the speed of light, he not only concluded that light must be a form of electromagnetic wave, but predicted that 'other radiations' could be 'propagated through the electromagnetic field according to electromagnetic laws' (see page 126). It was these other radiations, now called radio waves, that Hertz discovered in the mid 1880s.

Following in the footsteps of Michael Faraday (see page 99), Hertz had been carrying out experiments involving induction using a pair of Reiss spirals – coils of wire wound like a solenoid with their ends connected to metal balls separated by a small gap. He noticed that when electricity from a storage device known as a Leyden jar was discharged into one of these coils (via the metal balls), a spark jumped across the gap in the other coil, even though the two spirals were not physically connected and were not – as in Faraday's coils – wrapped around a shared iron rod. Some influence had travelled through the space between the spirals to make the second one spark in response to the activity of the first one.

In order to investigate the phenomenon, Hertz scaled up his experiment. He used an electric generator and an induction coil to make sparks not in a Reiss

Illustration of the apparatus used by Heinrich Rudolf Hertz (1857–1894) to discover radio.

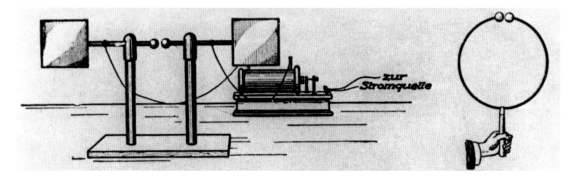

spiral but across a gap in the middle of a piece of wire two metres long, ending in spheres 4 millimetres in diameter. The long wire acted as a radiator for electromagnetic energy (in modern language, the antenna of a transmitter). The electric current associated with the sparks bounced to and fro along the wires. Various pieces of equipment were used to detect the waves being radiated by the antenna, the simplest being a piece of copper wire, 1 millimetre thick, bent into an almost complete circle 7.5 centimetres across. There was a small brass sphere on one end of the wire; the other end was pointed, nearly touching the sphere. The adjustable gap between the point and the sphere was typically a few hundredths of a millimetre. Sparks leaping across the gap in the transmitter produced corresponding sparks in the receiver, and by moving the receiver to different distances from the transmitter, over a range of a few metres, and measuring the strength of the response, Hertz could measure properties of the waves being radiated. He also carried out experiments in which a zinc plate was used as a reflector to generate standing waves for study.

In a series of experiments carried out between 1886 and 1889, Hertz found that the waves were four metres long, twice the length of the transmitting antenna, and travelled at the speed of light. He showed that these waves they could be reflected and refracted like light, and focused by concave reflectors. This was a dramatic confirmation of Maxwell's theory of electromagnetism, since visible light has wavelengths in the range 400 nanometres to 700 nanometres, where a nanometre is a billionth of a metre. The Hertzian waves were roughly ten million times bigger than light waves, but obeyed the same rules.

All of the discoveries were published in a series of papers in the journal *Annalen der Physik*, before being in presented Hertz's book, *Untersuchungen Ueber Die Ausbreitung Der Elektrischen Kraft* (*Investigations on the Propagation of Electrical Energy*), in 1892.

The book is now considered a classic, but Hertz had no idea of the practical implications of his discovery. He told a student that:

'It's of no use whatsoever; this is just an experiment that proves Maestro Maxwell was right – we just have these mysterious electromagnetic waves that we cannot see with the naked eye. But they are there.'

'What next?' asked the student.

Hertz replied,

'Nothing, I guess.'[25]

He was wrong. The waves he discovered – now called radio waves – revolutionized communications in the twentieth century, laying the foundations of radio, radar, TV and the wireless society on which we now depend. Appropriately, the modern unit of frequency (cycles per second) is named the Hertz (Hz) in his honour. In those units, the waves investigated by Hertz had a frequency of a few hundred Megahertz (MHz), generated by a half-wave dipole antenna.

NOBLE GASES AND A NOBLE LORD

Sometimes, the fruits of an experiment are a long time ripening – often, because the technology to investigate further has to be developed in order to gain a better understanding of what is going on. That is what happened after Henry Cavendish made a mysterious discovery in the mid-1780s.

Cavendish had been investigating the properties of what he called 'dephlogisticated air' (oxygen) and 'phlogisticated air' (nitrogen). By making sparks pass through a mixture of these gases, he was able to make what we now know as various oxides of nitrogen. But while carrying out these experiments he noticed something odd. If he started with a sample of air from the atmosphere (which is indeed mostly a mixture of nitrogen and oxygen), even after he had removed all traces of the two gases and all chemical activity had stopped, a small bubble of gas was left behind. This, he noted, was 'certainly not more than 1 / 125 of the bulk of the phlogisticated air', which shows what a skilled experimenter he was, but he had no idea what it could be.

These things stood for more than a century. Then, John William Strutt (the third Baron Rayleigh) was carrying out some accurate experiments to measure the density of different gases, as part of a project to measure atomic weights. Although Rayleigh had been a Professor in Cambridge, by that time he was working at his private laboratory in Terling, in Essex. In 1892 he noticed that the density of nitrogen extracted from the air was slightly greater than the density of nitrogen obtained by breaking down ammonia (NH_3) into its component elements. The natural explanation was that there was some impurity in the nitrogen obtained from the air, and Rayleigh urged his colleagues to investigate this.

One of those colleagues, William Ramsay, was working at University College, London, when he attended a lecture given by Rayleigh on 19 April 1894, where the puzzle was highlighted. After talking to Rayleigh, he followed the suggestion up, and by August that year had identified the impurity as a heavy gas, which refused to react with anything chemically, so he called it 'argon' (from the Greek word for lazy). It makes up 0.93 per cent of the Earth's atmosphere, and was the first known example of the so-called inert gases, sometimes known as 'noble' gases because they hold themselves aloof from chemical interactions. But as Rayleigh later pointed out, 'Argon must not be deemed rare. A large hall may easily contain a greater weight of it than a man can carry.'[26]

In his presidential address to the Royal Society at the beginning of 1895, Lord Kelvin referred to the discovery as the greatest scientific event of the preceding year. Ramsay and Rayleigh published a joint paper on the discovery later in 1895, and Ramsay went on to discover other inert gases, now known as helium, neon, krypton, xenon, and radon.

Henry Cavendish
(1731–1810).

It was only later that Ramsay realized why Rayleigh's lecture had struck a chord. In 1904, in the talk Ramsay gave when he received the Nobel Prize in Chemistry for his work, he said: 'I must have read the well-known account of Cavendish's classical experiment on the combination of the nitrogen and the oxygen of the air at that date; for in my copy of Cavendish's life, published by the Cavendish Society in 1849, opposite his statement that on passing electric sparks through a mixture of nitrogen with excess of oxygen, he had obtained a small residue, amounting to not more than 1/125th of the whole, I find that I had written the words 'look into this '. It must have been the latent memory of this circumstance which led me, in 1894, to suggest to Lord Rayleigh a reason for the high density which he had found for 'atmospheric nitrogen.'[27]

Ramsay's Nobel Prize was for chemistry; the same year, Rayleigh received the Nobel Prize for physics, for what was essentially the same work. Rayleigh's

William Ramsay (1852–1916) in his laboratory.

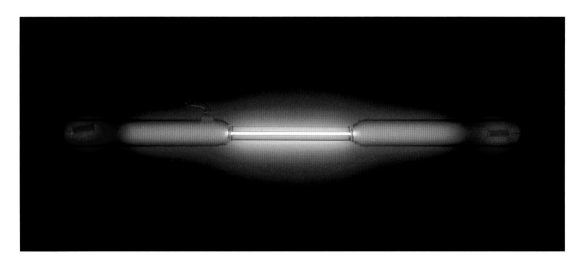

Electric discharge tube containing argon gas glowing with the characteristic colour of the argon spectrum.

citation read 'for his investigations of the densities of the most important gases and for his discovery of argon in connection with these studies', while Ramsay's read 'in recognition of his services in the discovery of the inert gaseous elements in air, and his determination of their place in the periodic system'. But the second citation highlights the significance of the discovery, and why it was worthy of such recognition.

The discovery of the inert gases was fundamentally important for the developing understanding of atomic structure. These gases formed a category (or group) of their own in the periodic table of the elements developed by Dmitri Mendeleyev. In the twentieth century, when Niels Bohr developed his explanation of chemistry in terms of atomic structure, the inertness of these gases was explained in terms of the distribution of electrons in their atoms, with the outermost layer, or shell, of electrons being in a stable configuration that does not allow the electrons to link up with other atoms.

N⁰· 55   THE BIRTH OF BIOCHEMISTRY

Even at the end of he nineteenth century, there were still respectable scientists (including Louis Pasteur) who argued that there was something special about the chemistry of life, and that some 'life force' was involved in what were called vitalistic processes. The final refutation of this idea came in 1897, from the work of the German Eduard Buchner.

Buchner decided to tackle the question of fermentation, which divided opinion at the time. Fermentation is a process that takes place without using oxygen in which living cells convert foods such as sugar into simpler compounds, such as alcohol and carbon dioxide, and the energy that powers

**Eduard Buchner
(1860–1917).**

the cells is released. Fermentation is a common process in living things, but Buchner studied the simple process of alcohol production, which involves yeast. Fermentation was seen by many to be a physiological act inseparably linked with the life processes of yeast, although Friedrich Wöhler (see page 99) had poked fun at the vitalism idea, as far back as 1839, by satirically summarizing the argument as: 'In a word, these infusoria gobble sugar, and discharge ethyl alcohol from the intestine and carbon dioxide from the urinary organs.' Yeast is indeed a living organism, and yeast was essential for the process. But was this because the yeast cells were alive, or because they contained some chemical substance that encouraged (catalysed) the conversion of sugar into alcohol and carbon dioxide without any need for vitalism? Could fermentation occur without the presence of living yeast cells?

The only way to find out was by experiment. Buchner had long been interested in the problem, and when he was appointed as Professor Extraordinary for Analytical and Pharmaceutical Chemistry at the University of Tübingen in 1896 he was able to carry out the work on a large scale in a fully equipped laboratory. By 9 January 1897 he was ready to send his key scientific paper, *Über alkoholische Gärung ohne Hefezellen* (*On Alcoholic Fermentation without Yeast Cells*), to the editors of the journal *Berichte der Deutschen Chemischen Gesellschaft*.

Yeast cells are like small bubbles filled with a semi-liquid substance, the protoplasm, surrounded by a comparatively firm cell membrane. To investigate the chemical composition of the cell contents, the experimenters had to remove the membrane by crushing it, because if they used any chemically active solvents or high temperatures this would alter the chemistry they wanted to study. And it was important that the process should be completed as quickly as possible, to minimise the possibility of any change while the experiment was going on.

So Buchner started with living yeast cells, but then treated them to a series of processes which killed them and reduced them to their constituent chemical parts by purely physical means. These included mixing dry yeast cells, grains of quartz sand, and a soft crumbly rock (rather like pumice), known as diatomite, and then grinding the mixture, including the yeast cells, with a pestle and mortar. The mixture became damp as the yeast cells ruptured and their contents were released. Then, the damp mixture, which resembled a thick dough, could be squeezed to extract a 'press juice' used in the experiments.

This process was very efficient. Half a litre of liquid could be obtained from 1,000 grams of yeast, providing plenty of material to work with. When sugar solution was added to freshly pressed yeast juice, a strong production of gas occurred; in containers where the juice had been mixed with concentrated sugar syrup and left for several hours there was active frothing of carbon dioxide bubbles, and a thick layer of foam formed, showing that fermentation was occurring. When sugar was dissolved in juice at blood heat, these effects were visible within fifteen miniutes. Careful investigations by Buchner and his colleagues established that carbon dioxide and alcohol were produced in exactly the same proportions as in fermentation with live yeast. But microscopic investigation revealed no living yeast cells in the extract.

In further studies, Buchner found that the key chemical substance that promotes the breakdown of sugars in this way is an enzyme, which he called zymase. Zymase is a protein that is manufactured inside yeast cells, so in that sense life is involved in the process of fermentation. As Buchner put it in his Nobel Lecture (he was awarded the chemistry prize in 1907 for his biochemical researches and his discovery of cell-free fermentation), 'the difference between enzymes and micro-organisms is clearly revealed when the latter are represented as the producers of the former, which we must conceive as complicated but inanimate chemical substances'. But the crucial point is that the chemistry carries on whether the yeast is alive or dead. Enzymes are crucial players in many biological processes, but it is now possible to synthesize enzymes chemically without biology being involved.

Froth produced by baker's or brewer's yeast, *Saccharomyces cerevisiae*, in the process of fermenting sugar.

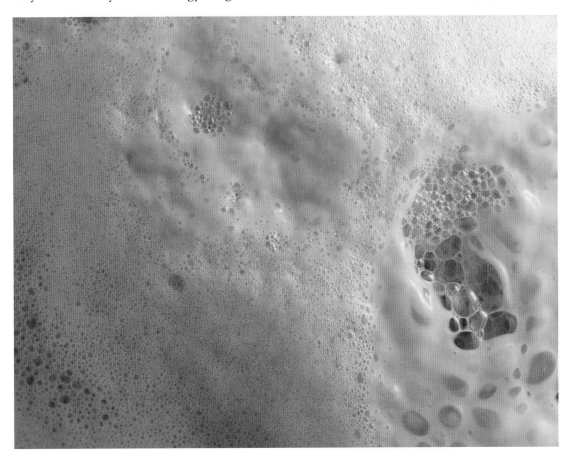

ENTER THE X-RAY

T he close link between experimental science and technology is particularly
powerfully demonstrated by the development of physics at the end of the
nineteenth century. One key piece of technology, the vacuum (or Crookes) tube,
produced a revolution in the understanding of the world of the very small, and
led to the birth of atomic physics.

In 1894, Philipp Lenard, working in Germany and following up experiments
carried out by Heinrich Hertz, studied the way cathode rays could pass through
a thin sheet of metal foil inside a vacuum tube. Because the rays passed
through the metal without leaving any detectable holes, he thought that they
must be waves, not particles. This idea would soon be proved wrong (see page
152), but in a separate development Wilhelm Röntgen, the professor of physics
at the University of Würzburg, decided to follow up Lenard's work to find out
if the cathode rays could penetrate the glass of the Crookes Tube itself.
Röntgen was already 50 years old when he carried out his key experiments in
November 1895 – an established scientist with a reputation as a skilled
experimenter, but still open to new ideas.

**Laboratory of
Wilhelm Konrad Röntgen
(1845–1923).**

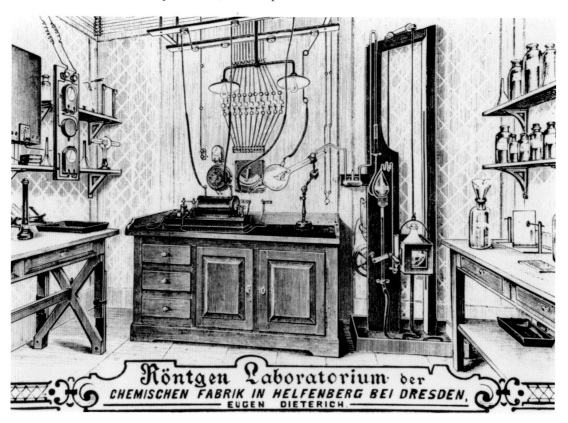

To see if cathode rays were passing through the glass of the tube and out into the laboratory, Röntgen set his apparatus up in a darkened room and covered the entire tube with a thin layer of black cardboard to stop any light from the glow inside escaping. He wanted to be sure his eyes were adapted to the dark and could pick up any flashes from the simple cathode-ray detector he had prepared. It was already known that a sheet of paper painted with barium platinocyanide would fluoresce when struck by cathode rays, so Röntgen set up such a screen in the dark lab to look out for the fluorescence. After carrying out some initial tests, Röntgen decided to repeat the experiment using a vacuum tube with a thicker glass wall. On 8 November 1895, after he had set up the new tube, and covered it with black cardboard, he put the barium platinocyanide screen to one side while he turned the vacuum tube on and the lab lights off, simply to make certain no light was escaping from the tube. To his surprise, he saw a faint shimmering light from the screen, which was several feet away from the tube and off to one side, not in the 'line of fire' of the cathode rays. Something else – something previously unknown – was making the screen glow when the tube was turned on.

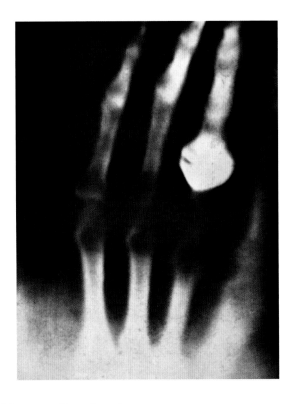

First X-ray photograph of a human being (1895). The picture was made by Roentgen and shows the hand of his wife, who is wearing a ring.

Over the next few weeks, Röntgen made a careful study of what he called 'X-rays' ('X' is traditionally the unknown quantity; in some countries they became known, to his embarrassment, as Röntgen-rays). He found that they were produced where the cathode rays hit the glass wall of the Crookes tube, and spread out in all directions. They travelled in straight lines, and were not deflected by electric or magnetic fields. But the most dramatic discovery was that they would penetrate many materials, including human flesh. Just two weeks after the discovery of X-rays, Röntgen used them to make the first X-ray photograph, of his wife's hand. It showed the bones of her fingers and her wedding ring, and it was published in the scientific paper in which he announced the discovery. Titled 'On a new kind of ray' ('Über eine neue Art von Strahlen'), it was written on 28 December 1895, and published early in 1896.

The discovery was sensational news, not least because of the publication of the X-ray photograph of Anna Röntgen's hand (on seeing the image, apparently she exclaimed 'I have seen my death!'). On 13 January 1896, Röntgen demonstrated the phenomenon to Emperor Wilhelm II, in Berlin. English translations of the paper were published in the journals *Nature* (on 23 January) and *Science* (on 14 February). X-rays were soon recognized as part of the electromagnetic spectrum as light, but with higher frequencies (shorter wavelengths).

The medical implications of the discovery were obvious, and X-rays soon became an important diagnostic tool, enabling doctors to see inside the human body without cutting it open. Before the end of 1896, the first radiology department opened in a Glasgow hospital, where they obtained pictures of a kidney stone and of a penny lodged in a child's throat. Less than two years after the discovery, X-rays were used for the first time in a battlefield context, to find bullets and broken bones inside patients during the Balkan War of 1897.

The first Nobel Prize in physics was awarded to Röntgen in 1901, 'in recognition of the extraordinary services he has rendered by the discovery of the remarkable rays subsequently named after him.'

Nᵒ· 57   ENTER THE ELECTRON

The electron was a discovery waiting to happen, and several experimenters converged on the discovery in the 1890s. The key contributor was the British physicist J. J. Thomson (always referred to by his initials, never his full name), working at the Cavendish Laboratory in Cambridge.

In 1894, Thomson found the first clue when he measured the speed of cathode rays passing through a Crookes Tube (see page 138) and found that it is much less than the speed of light. So cathode rays could not be a form of electromagnetic radiation, because Maxwell's equations tell us that all such radiation travels at the same speed. A year later, the Frenchman Jean Perrin

J. J. Thomson (1856–1940).

discovered that a beam of cathode rays could be bent sideways by a magnetic field, implying that the beam is made up of a stream of electrically charged particles; the direction of the deflection showed that the particles must carry a negative electric charge.

But what were the particles? Walter Kaufmann, working in Berlin, thought that they must be electrically charged atoms (now known as ions). He thought that cathode rays were atoms that had picked up negative electric charge from the cathode, and he measured the way cathode rays were deflected by electric and magnetic fields in vacuum tubes containing traces of different kinds of gas. He was able to work out the ratio of the charge of the particles to their mass (usually written as e/m),

and expected to get different values for the different gases, because they had different atomic weights,. But he always got the same value for e/m.

Thomson also measured e/m for cathode rays, but unlike Kaufmann he expected to get the same value all the time, because from the start he thought that cathode rays were streams of identical particles being emitted from the cathode. So he used tubes with as pure a vacuum as possible (the best vacuum tubes in the world at the time) to avoid contamination, and devised a clever technique in which the beam was simultaneously being pushed one way by a magnetic field and the opposite way by an electric field. By adjusting the strengths of these fields, he could end up with a beam travelling in a straight line, and the strengths of the fields needed to achieve this enabled him to work out e/m.

Although at this stage Thomson had a value only for the ratio, not for the charge and mass separately, he could compare this with the value he got for similar experiments he carried out with electrically charged atoms (ions) of the lightest element, hydrogen. This told him that either the mass of the particles of cathode rays was very small, or the charge was very large, or some combination of these effects. This raised the possibility that the particles involved, which he called corpuscles, were smaller than atoms, and might be parts of atoms that had escaped or been knocked off. In a lecture to the Royal Institution on 30 April 1897, Thomson said, 'the assumption of a state of

Apparatus used by Thomson which lead him to the discovery of the electron.

matter more finely divided than the atom is a somewhat startling one'. He later recalled that at least one member of the audience thought that he had been pulling their leg; the idea of the atom not being indestructible was that outrageous in the 1890s.

Thomson's 'corpuscles' were soon dubbed 'electrons', and with the benefit of hindsight 1897 is now usually regarded as the year of the 'discovery' of the electron. But it was another two years before Thomson was able to measure the electric charge itself, in experiments involving ultraviolet light falling on metals and triggering the emission of negatively charged particles – the photoelectric effect. He measured e/m for these particles and showed they were identical to his cathode ray corpuscles, then measured their charge by monitoring water droplets charged by these particles moving in electric fields. Combining these measurements with the measurements of e/m made it possible to work out the mass of the individual corpuscles/electrons, which turned out to be just one two-thousandth of the mass of a single hydrogen atom. In the *Philosophical Magazine*, Thomson wrote in December 1899: 'On this view, electrification essentially involves the splitting up of the atom, a part of the mass of the atom getting free and becoming detached from the original atom.'[28] The atom, it was clear, is not indivisible, and this marked the real moment of the discovery of the electron.

NO. 58   RADIOACTIVITY REVEALED

The discovery of X-rays (see page 150) almost immediately led to another equally profound discovery. In January 1896, when news of Röntgen's work spread through the scientific community, Henri Becquerel was professor of physics at the French Museum of Natural History, in Paris. His father and grandfather had each held the post before him, and there was a family tradition of studying luminescent phenomena, especially the phosphorescence of crystals that glow in the dark. These crystals have to be exposed to sunlight in order to become energized, then they glow for a while before the glow fades as the energy is used up. Becquerel wondered if the glowing crystals might also produce X-rays, and devised a simple experiment to find out, using the wealth of phosphorescent material that had accumulated in the lab since his grandfather's time.

He 'charged' a dish of phosphorescent salts by exposing them to sunlight, then stood the dish on top of a photographic plate that had been wrapped in two sheets of thick, black paper, so that no light could penetrate. In some tests, he put a coin between the dish and the photographic plate; in others he used a piece of metal in the shape of a cross. As he had hoped,

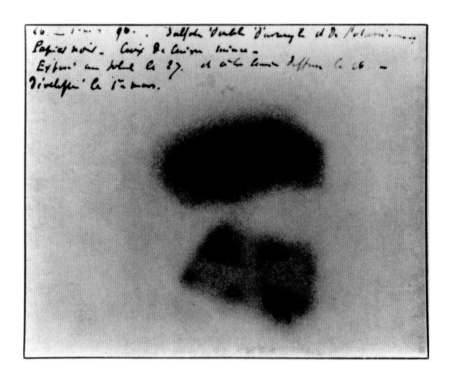

The photograph that led Henri Becquerel (1852–1908) to the discovery of radioactivity. His comments are written on the photograph.

when the photographic plates were developed they had been fogged, just as if they had been exposed to light, but the outlines of the metal objects showed up clearly, because X-rays could not penetrate the metal. It seemed that X-rays could be produced by the action of sunlight on phosphorescent salts, and this news spread hot on the heels of the news about X-rays.

But at the end of February that year, Becquerel accidentally discovered something astonishing. He had prepared another experiment, using a dish containing uranium salts, a cross made of copper, and the usual wrapped photographic plate. Because the weather was overcast, he left it in a cupboard until, on 1 March 1896, either on a whim or deliberately as a 'control' experiment, he developed the plate anyway. There was the clear image of the cross. Even though the uranium salts had not been charged up by sunlight and were not glowing, they had produced what he assumed must be X-rays. In further tests, Becquerel found, as he would later recall in his Nobel Lecture: '[…] that all uranium salts, whatever their origin, emitted radiation of the same type, that this property was an atomic property connected with the element uranium, and that metallic uranium was about three and a half times as active as the salt used in the first experiments.'[29]

The discovery did not make a very big splash, because everyone thought that all Becquerel had found was another way to make X-rays. But in 1897 a young Polish woman who had moved to Paris and married, to become Marie Curie, set out to work on 'uranium rays' as her PhD project. It was Marie Curie who gave the phenomenon the name 'radioactivity'. In February 1898,

Part of the apparatus used by the Curies and their assistants in the period 1898–1902 in a shed in Paris, France, to isolate the radioactive element radium. This involved processing tons of pitchblende, a uranium-rich ore.

while working under extremely difficult conditions (literally in a leaky shed, because women were not then allowed to enter the laboratories at the *École Normale Supérieure*, where she taught), she found that pitchblende, the ore from which uranium is derived, is four times more radioactive than uranium. This meant that it must contain another previously unknown and highly radioactive element. The discovery was announced in a paper presented to the *Académie des Sciences* on 12 April 1898; in it she said, 'The fact is very remarkable, and leads to the belief that these minerals may contain an element which is much more active than uranium.' It was such a dramatic discovery that her husband, Pierre Curie (see page 139), dropped his own work and helped her track down this new element. In fact, they found two – one they called polonium, as a tribute to her homeland, the other they called radium.

However, it took until March 1902 to extract just one-tenth of a gram of radium from several tonnes of pitchblende. By then, Becquerel (among others) had found that 'uranium rays' could be deflected by a magnetic field, so they could not be the same as X-rays and must be made up of a stream of electrically charged particles. These are now known as alpha particles, and are identical to nuclei of helium atoms (helium atoms with the electrons removed). Marie Curie's PhD was duly awarded in 1903, the same year that the Nobel Prize in Physics was awarded jointly to Becquerel, 'in recognition of the extraordinary services he has rendered by his discovery of spontaneous radioactivity', and to Pierre and Marie, 'in recognition of the extraordinary services they have rendered by their joint researches on the radiation phenomena discovered by Professor Henri Becquerel'.

Nᴼ· 59 KNOCKING ELECTRONS WITH LIGHT

During his work with what we now call radio waves (see page 143), in 1887 Heinrich Hertz noticed that it was easier to make sparks jump across the gap between his electrodes if they were radiated with ultraviolet light. He had put his apparatus in a darkened box, so that he could see the spark better, with a glass window to look through. But he noticed that the sparks would not jump across as big a gap as when the electrodes were not in the box. When the glass window was taken away, leaving a hole, the sparks were able to jump as before. By trying windows made of different substances, he worked out that the glass was absorbing ultraviolet light, and that this was responsible for the effect.

Although Hertz published his results, he did not suggest any explanation for the effect, and he did not carry out further experiments to investigate it. Other people did investigate the phenomenon (in particular, the Russian physicist Aleksandr Stoletov), but the key experiment that led to an understanding of the phenomenon was carried out by Philipp Lenard, in 1902. Lenard had worked as an assistant to Hertz in Bonn, but by then he was established as a professor in his own right, at the University of Kiel. He was carrying out a major investigation of cathode rays (electrons) and wanted to find out if the effect discovered by Hertz was a result of ultraviolet light releasing electrons from a metal surface. By shining ultraviolet light on a clean metal plate in a vacuum tube he was able to produce cathode rays, which could be manipulated by magnetic and electric fields in the usual way. But the particles leaving the metal surface travelled so slowly that they could be stopped and made to fall back by a small electric potential. The speed of the particles could be measured by the strength of the electric field needed to make this happen, and this led to Lenard's most profound discovery: 'I have also found that the velocity is independent of the ultraviolet light intensity.'[30]

Philipp Lenard (1862–1947).

Photoelectric effect.
1) Blue light (strictly speaking, ultraviolet) shone on a metal sheet makes it eject electrons.
2) With less blue light, fewer electrons are ejected, but each has the same energy as before.
3) Using a red light, no electrons are ejected.

It would have been natural to expect that a brighter light would make the electrons move faster, with more energy. But increasing the brightness only produces more electrons, each with the same energy. Further experiments showed that changing the frequency (or wavelength) of the light does affect the energy of the electrons. Light with a higher frequency (shorter wavelength) produces electrons with more energy; light with lower frequency (longer wavelength) produces electrons with less energy.

In 1905, the same year that Lenard received his Nobel Prize, Albert Einstein found the explanation. He suggested that light exists in the form of packets of energy, or quanta, which became known as photons. On this picture, electrons produced by the photoelectric effect are ejected when a single photon strikes a single atom. Each photon striking an atom gives up all of its energy to the ejected electron. What we think of as higher frequency light is made up of more energetic photons, so when a higher frequency photon hits an atom in a metal surface, a single electron is released with the same high energy as the incoming photon. A brighter light just carries more photons with the same energy, so releases more electrons with the same energy. Similarly, for light of a particular frequency, a fainter light carries fewer photons, so it releases fewer electrons, but still all with the same energy.

Einstein's suggestion that light could somehow be both a particle and a wave was received sceptically. After all, the evidence of the double-slit experiment (see page 79) still stood. The American physicist Robert Millikan was so outraged that he spent ten years carrying out a series of difficult experiments aimed at proving Einstein was wrong, only to conclude that light quanta were real. It is worth quoting his own comments: 'After ten years of testing and changing and learning and sometimes blundering, all efforts being directed from the first toward the accurate experimental measurement of the energies of emission of photoelectrons, now as a function of temperature, now of wavelength, now of material (contact e.m.f. relations), this work resulted, contrary to my own expectation, in the first direct experimental proof, in 1914, of the exact validity, within narrow limits of experimental error, of the Einstein equation.'[31]

Einstein's Nobel Prize (the 1921 physics prize, but held over until 1922) was awarded specifically 'for his services to Theoretical Physics, and especially for his discovery of the law of the photoelectric effect'. In 1923, Millikan received the physics prize 'for his work on the elementary charge of electricity and on the photoelectric effect'. As these awards show, the 'discovery' of photons was a key step on the road to quantum theory, where the concept of wave-particle duality (see page 181) proved crucial in developing an understanding of the behaviour of atoms and subatomic particles.

A PAVLOVIAN RESPONSE

The work for which Ivan Pavlov become famous provides a classic example of a discovery being made by a combination of luck and the scientific approach to observation and experiment. Pavlov was a Russian physiologist, who was interested in the mechanism of digestion. At the end of the nineteenth and beginning of the twentieth century he was studying how this worked in dogs. He had doubts about the value of the then new studies of psychiatry and psychology; but by chance he made his name through experiments in psychology.

As part of his study of digestion, Pavlov and his assistant, Ivan Tolochinov, were carrying out experiments on dogs. Using delicate surgical techniques, the digestive juices and products of the salivary glands were diverted so that some of these secretions could be collected outside the bodies of the dogs without otherwise affecting the animals. The point was to carry out the studies on living animals going about their more or less normal lives, instead of relying on

Former home of Ivan Pavlov,
now a museum in Ryazan.

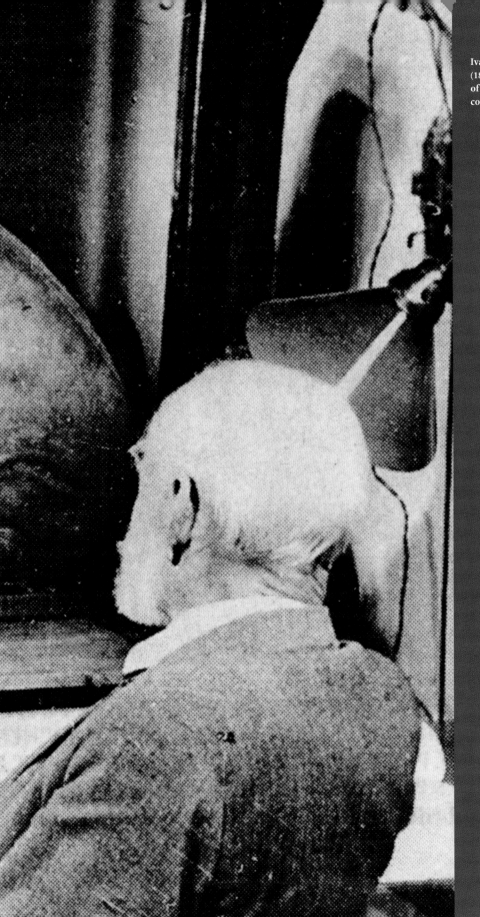

Ivan Petrovich Pavlov (1849–1936) observing one of the dogs on which he conducted his experiments.

dissections of dead animals or invasive surgical procedures on living but immobilized animals (who subsequently died). As Pavlov put it: 'It is perfectly clear that successful study of the digestive process, as of any other function of the organism, depends to a considerable degree on whether we succeed in finding the nearest and most convenient point of view on the process under observation and in removing all intervening processes between the phenomena under observation and the observer.'[32]

Pavlov was extremely proud of the care taken to ensure that the animals were pain free and well looked after. He went on: 'Our healthy and happy animals did their laboratory work with real gusto; they always eagerly moved from their cages to the laboratory and readily jumped onto the tables where our experiments and observations were conducted.'[33]

This made it possible to measure the flow of digestive juices under different conditions, and according to what food the dogs were fed. It was no surprise to find that the dogs salivated at the sight of food. But the food was always brought to them by the same technician, and Tolochinov noticed that after a time the dogs salivated whenever they saw the technician, even if they were not being fed.

Intrigued, Pavlov organized a series of experiments in which different kinds of stimulus – such as bells and buzzers, flashing lights, and ticking metronomes – were presented to the dogs at the time the food was offered. In the classic example, he rang a bell just before giving the dogs their food. Soon, the dogs began to salivate every time the bell was rung, even when no food was offered. He called this a conditional response (the term was later modified to conditioned response), and found that the response was stronger if the food had been served more quickly after the bell was rung during the initial conditioning. But when he stopped serving food at the time the bell was rung, the conditioned reflex (now sometimes called a Pavlovian response) wore off, and the dogs stopped salivating at the sound of the bell. Pavlov called the bell a conditional (conditioned) stimulus, because the dogs were conditioned to respond to it, by contrast with food, which was an unconditional (unconditioned) stimulus producing an unconditional (unconditioned) response.

Tolochinov described the experiments at a scientific meeting in Helsinki in 1903, and Pavlov presented them to a meeting in Madrid later that same year. As news of the discovery spread, it had a big influence on the way people thought about themselves, as it was realized that people could be 'conditioned' by experience to respond in a certain way to specific stimuli. Aldous Huxley presented a dystopian view of the implications of this in his 1932 novel, *Brave New World*.

Later experiments showed that conditioned reflexes originate in the cerebral cortex (they are literally learned), and stimulated further research into the way the brain works. Although Pavlov received the Nobel Prize in Physiology or Medicine in 1904, it was too soon for the importance of his work on conditioning to be recognized, and it was specifically 'in recognition of his work on the physiology of digestion'.

xperiments' are not always triggered by people, but scientists can still make good use of them by careful observation and analysis. The discovery of the Earth's core is a classic example.

Emil Wiechert, a German geophysicist, realized in the middle of the 1890s that the interior of the Earth must contain a core much more dense than the surface layers. Thanks to the Cavendish experiment (see page 64), he knew the mass and overall density of the whole Earth, and it is straightforward to measure the density of rocks at the surface. He worked out a detailed model for the inner structure of the Earth to fit the observations, and presented an early version in a lecture in Königsberg in 1896, followed the next year by a full description. He estimated that the Earth has an iron core with a radius 0.779 times the radius of the Earth and a density of 8.21 grams per cubic centimetre, while the outer layer (the mantle) has a density of only 3.2 grams per cubic centimetre. These are not hopelessly different from modern measurements (see below).

Wiechert realized that this idea could be tested by studying earthquake waves, which travel from their point of origin through the interior of the Earth and can be detected on the other side of the planet. This was the necessary natural

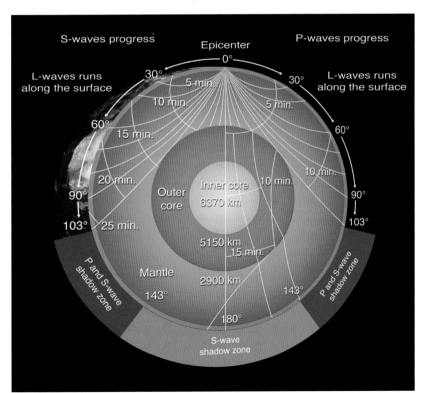

Computer-generated illustration of a section through the Earth, showing the time taken for different types of seismic waves to propagate around the globe.

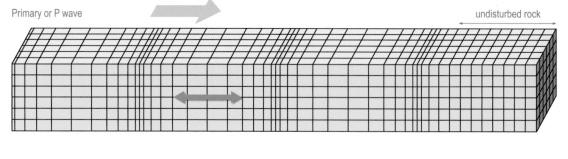

Primary or P wave

undisturbed rock

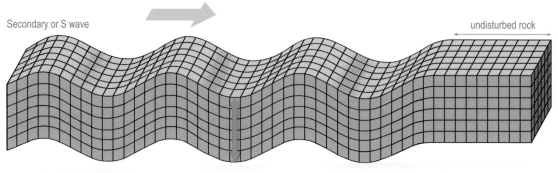

Secondary or S wave

undisturbed rock

The two types of body wave produced by an earthquake. The Primary wave (P wave, top) is a longitudinal wave that causes ground displacement (blue arrows) in the same direction as the wave's movement (yellow arrow). P waves are analogous to sound waves in air. Secondary waves (S waves, bottom) have displacement at right angles (blue arrows) to the direction of movement. They are analogous to ripples on a pond.

experiment to test his theoretical model. Coincidentally, in 1897, the same year that Wiechert presented his detailed model, a major Earthquake in India set the British geologist Richard Oldham on the path to making that test.

Oldham was born in India, where his father was a geologist studying earthquakes. In his turn, he joined the Geological Survey of India and made a major study of the magnitude 8.1 Assam earthquake of 1897, which was felt over an area of 250,000 square miles. From this work he discovered that there are three different kinds of seismic waves, which travel at different speeds through the earth. P-waves (the P stands for either Primary or Pressure) travel through the deep Earth and are the first to arrive at a distant seismograph. They are 'push-pull' waves, like sound waves in air. The second waves to arrive at the seismograph are known as S-waves (the S stands for Secondary or Shear), which also move through the deep Earth, but with a wave motion at right angles to the direction of wave movement, like ripples on a pond. There are also surface waves, which travel along the surface of the Earth with devastating but local effects.

Oldham left India in 1903 to settle in Britain, where he used this new insight to analyse seismic data from around the world for thousands of earthquakes (thousands of natural experiments). He studied the way waves are refracted and change speeds as they pass through the interior of the Earth, and found that earthquake waves initially increase in speed as they move into the Earth, but that at a certain depth they suddenly slow down. This showed that the wave had reached the boundary of the core. His results were published in 1906, the year of the great San Francisco earthquake. Oldham was awarded the Lyell medal of the Geological Society of London in 1908.

Wiechert, who had seen iron meteorites, had suggested that the Earth might be like a giant meteorite with a core of nickel-iron metal which had settled to the centre. Oldham's measurements of the change in speed of earthquake waves within the Earth implied that the core must actually be liquid. It was only in 1936 that the Dane Inge Lehmann discovered that there is a small, solid inner core which reflects seismic waves. Modern measurements place the top of the outer core at a depth of 2,890 kilometres below the Earth's surface, while the inner core begins at a depth of 5,150 kilometres. The radius of the Earth is 6,360 kilometres. Because volume goes as the cube of the radius, this means that the whole core occupies 16 per cent of the volume of the Earth, and the inner core just under 1 per cent of the volume (the core has roughly half the radius of the Earth and therefore roughly one eighth of the volume, because 2 cubed is 8). The densities range from 9.9 to 12.2 grams per cubic centimetre in the outer core, and slightly higher, perhaps 13 grams per cubic centimetre, in the inner core. The inner core may be crystalline, although it is very hard to test this idea. Swirling fluid currents in the outer core are probably responsible for generating the Earth's magnetic field.

Nº 62 INSIDE THE ATOM

rnest Rutherford is a very unusual example of an experimental physicist who made his most important discovery *after* he had received the Nobel Prize for something else. The prize, awarded in 1908, was actually the Nobel Prize in Chemistry, for his work on what was called 'the disintegration of the elements'. Following up Becquerel's discovery of radioactivity (see page 154), Rutherford had shown that the process involves an atom spitting out a part of itself and being transformed into another kind of atom. Along the way, he identified two kinds of radiation, which he called alpha rays and beta rays; alpha rays were soon identified as helium atoms lacking their electrons, and beta rays as electrons. He later found and named a third kind of atomic radiation, gamma rays, which turned out to be like X-rays but with even shorter wavelengths (higher frequency) and therefore carrying even more energy.

Not content with just identifying and classifying different kinds of atomic radiation, Rutherford realized that he could use these fast-moving particles, in particular alpha particles, to probe the structure of matter. At the beginning of the twentieth century, although electrons had been identified as parts of an atom (see page 152), nobody was sure how they were arranged inside an atom, or

Ernest Rutherford (1871–1937), left, and Hans Geiger (1882–1945), right, in their laboratory at Manchester University in about 1908. They are seen with the equipment they used to detect alpha particles from a radioactive source.

where the positive charge needed to balance the electrons' negative charge was located. The most popular idea was proposed in 1904 by J. J. Thomson, who envisaged the atom as a cloud of positive charge in which negative electrons were embedded. This was sometimes known as the 'plum pudding' model, but a better analogy would be with a watermelon, where the flesh of the melon is the positive charge and the pips are electrons.

In 1909, Rutherford was Professor of Physics at the University of Manchester, where he devised an experiment that was carried out, under his direction, by Hans Geiger and Ernest Marsden. Alpha particles produced by natural radioactivity were fired towards a target in the form of a thin sheet of gold foil. Using a detector that could be moved to different places around the foil (the forerunner of the detector named after Geiger), they planned to find out how the alpha rays were affected as they passed through the foil. They anticipated that the beam of particles might be deflected slightly, like a beam of light passing from air into a glass prism. But to their surprise they found that most of the alpha particles went straight through the foil as if it were not there, while some of them were being deflected at large angles. Even when they moved the detector round to the front of the foil, on the same side that the alpha particles were coming from, some particles were detected, having been bounced back off the foil almost in the direction they came from. It was as unexpected as if you threw a brick at a sheet of wet tissue paper hung up on a line and it bounced back to hit you in the face.

Rutherford described it as 'quite the most incredible event that has ever happened to me in my life', but soon worked out what was going on. The atom must actually be composed of a very small, positively charged nucleus (a term first used in this context by Rutherford, in 1912), surrounded by a cloud of electrons. An alpha particle has eight thousand times as much mass as an individual electron, and most alpha particles in the experiment shoot through the electron clouds almost entirely unaffected. But alpha particles, like the nuclei of atoms, have positive charge, and if one happens to pass close by a nucleus it is deflected by a large angle, because like charges repel each other. On rare occasions, an alpha particle heading straight towards a nucleus is bounced back the way it came, because a nucleus of gold has 49 times the mass of an alpha particle and cannot be brushed aside.

By counting the number of particles deflected at different angles in a long run of experiments and applying an appropriate statistical analysis, Rutherford was even able to work out roughly the size of the nucleus. Using more modern measurements, a typical nucleus 10^{-13} centimetres across is

Rutherford's particle-scattering experiment. Some particles pass through the sheet of gold foil without being deflected, others are deflected at a tiny angle, and just a few are reflected at a very large angle.

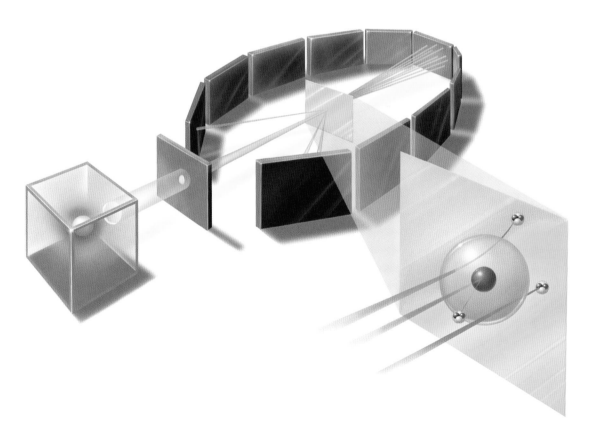

surrounded by a cloud of electrons 10^{-8} centimetres across. The proportions are roughly those of a grain of sand (the nucleus) placed in the centre of the Albert Hall (the electron cloud). Atoms, Rutherford had discovered, are mostly empty space, in particle terms, filled with a web of electric and magnetic fields linking tiny particles. The nucleus itself, later experiments showed, is made up of positively charged protons (hydrogen nuclei), one for each electron in the cloud of an atom, and electrically neutral particles similar to protons, called neutrons. Alpha particles are nuclei of helium, made up of two protons and two neutrons.

Nº. 63 A RULER FOR THE UNIVERSE

You might think that measuring the size of the Universe is an impossible task, beyond the scope of any experiment or Earth-bound observations. But between 1908 and 1912 an astronomer working at Harvard College Observatory, Henrietta Swan Leavitt, discovered a way to measure the distances to certain key stars, providing a ruler to determine distances across the Universe.

Leavitt was carrying out a survey of stars which vary in brightness. These variations may happen because the star actually does get brighter and dimmer as time passes, sometimes with a regular cycle, or because it is in a binary pair, where one star gets eclipsed at regular intervals as the other star passes in front of it. By this time, astronomers no longer had to rely on their unaided eyes, and she worked with photographic plates containing images of thousands of stars in the southern skies observed at different times.

After many hours of painstaking analysis, Leavitt noticed a pattern in the behaviour of a certain kind of star in a large concentration of stars, known as the Small Magellanic Cloud (SMC). The family of stars known as Cepheids had an overall pattern of behaviour in which the brighter stars (taking the average brightness over their whole cycle of brightening and dimming) went through the cycle more slowly than the fainter Cepheids. By 1912, she was able to work out this 'period-luminosity relationship' in terms of a mathematical formula, based on measurements of the variation of 25 Cepheids in the SMC. She realized that the pattern showed up because the SMC is a cloud of stars so far away from us that the separation of the stars in the cloud is small compared with the distance from the cloud to us. So light from all the stars gets

Henrietta Swan Leavitt (1868–1921).

dimmed by the same amount on its way to our telescopes. For nearby stars, it is hard to tell if one star looks brighter than another because it really is brighter, or because it is closer. But as the variables in the SMC 'are probably at nearly the same distance from the Earth,' she wrote in 1912, 'their periods are apparently associated with their actual emission of light, as determined by their mass, density, and surface brightness.'

Leavitt found that, for example, a Cepheid with a period of three days is one-sixth as bright as a Cepheid with a period of thirty days. But these were all relative measurements. Astronomers still needed to measure the actual distances to a few nearby Cepheids, so that they could work out their actual (absolute) brightnesses, and use these to calibrate the distance scale. Once they knew the distance to one Cepheid, then they could work out its absolute brightness. Using the relationship Leavitt had discovered, they could then work out the absolute brightnesses of other Cepheids, so their apparent brightnesses on the sky would reveal their distances.

The first such measurements, based on various techniques which only work for nearby stars, were made in 1913, but they were somewhat inaccurate. Modern measurements imply a distance to the SMC of 170,000 light years, so that even if two Cepheids in the cloud are 1,000 light years apart this represents only 0.6 per cent of their distance from us.

In the 1920s, Cepheids were used first to measure distances across our Milky Way Galaxy and determine its size, then to measure distances to what

Two ways to measure distances across the universe. On the left is the 'standard candle' method. Supernovas have a known brightness. The further you are from a supernova, the fainter it appears (as does a candle). Therefore by measuring how bright a supernova appears, its distance can be determined. On the right is the 'standard ruler' method. Galaxies look smaller if they are further away. Both methods are calibrated using Cepheid variable stars.

turned out to be other galaxies beyond the Milky Way, showing that the Milky Way is not unique but just an ordinary island in space, one of hundreds of billions of such islands in the observable Universe. It was by using Cepheid distances for galaxies that Edwin Hubble was able to work out the relationship between the redshift in the light from a galaxy and its distance, leading to the discovery that the Universe is expanding, and to the idea, put forward by Georges Lemaître (who independently came up with the redshift-distance relationship), of the Big Bang.

Over the decades since the work of these pioneers, other cosmic distance indicators have been developed, most notably certain kinds of supernovas, stars which all explode with the same absolute brightness. But we only know the distances to the archetypal supernovas used to calibrate this part of the distance scale because they occur in galaxies whose distances have been measured using Cepheids. This is so important that developments from Leavitt's work continue today. The Hipparcos satellite, launched in 1989 by the European Space Agency, measured accurate distances to more than 100,000 stars, including 273 Cepheids, using geometric techniques. This provided the most precise calibration of the cosmic distance scale in the twentieth century, but a follow-up mission, dubbed Gaia, was launched in 2013 and should provide even more accurate measurements by 2020. More than a hundred years after Leavitt's discovery, developments from her work continue.

N°. 64   THE DISCOVERY OF NUCLEIC ACIDS

ome discoveries are hard to pin down to one experiment, or one person. The 'discovery' of the life molecules RNA and DNA involved many people and many experiments, over nearly a hundred years. But the person who actually gave these molecules their names was Phoebus Levene, a Russian-born American working at the Rockefeller Institute of Medical Research.

The story really begins, however, with experiments carried out by a Swiss biochemist, Friedrich Miescher, working at the University of Tübingen in the 1860s. By that time, it was known that proteins are the most important physical substances in the body, and Miescher wanted to identify the proteins involved in the chemistry of the cell – a key to the workings of life. He got his raw material from pus-soaked bandages obtained from a nearby surgical clinic, and isolated the white blood cells known as leucocytes from the pus. Living cells are like little bags of watery jelly, called cytoplasm, with a more compact central nucleus (the term Rutherford later borrowed to describe atomic structure). Miescher found that the watery jelly is indeed rich in proteins, but he was also able to remove this outer material and collect enough nuclei to

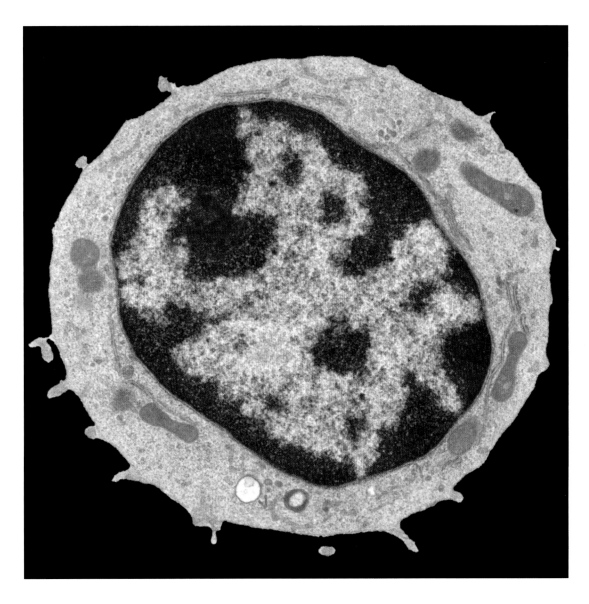

analyse them separately. He found that the nuclei were not made of protein, but of a different material, which he dubbed 'nuclein'. Just like other molecules of life, it contained a lot of carbon, hydrogen, oxygen, and nitrogen, but it also contained phosphorus, which is not found in any protein. He wrote, 'I think that the given analyses – as incomplete as they might be – show that we are not working with some random mixture, but … with a chemical individual or a mixture of very closely related entities.' But he was not able to work out the structure of the large nuclein molecules. In later work, he found that these molecules contain several acidic groups, and the term 'nucleic acid' began to be used to describe nuclein by the end of the 1880s.

Coloured transmission electron micrograph (TEM) showing the large central nucleus (brown) of a small lymphocyte (white blood cell). Surrounding this is the cytoplasm (light green), containing mitochondria (darker, solid green).

Phoebus Levene (1869–1940).

In the following years, analysis of nucleic acids showed that there are two types. One contains four sub-units, known as bases, with the names adenine, guanine, cytosine, and thymine – often referred to by their initials as A, G, C and T. The other nucleic acid contains a different base, uracil (U), instead of thymine.

This was the state of play in the second half of the first decade of the twentieth century, when Phoebus Levene began experimenting with nucleic acid derived from yeast cells. This contained about equal amounts of A, G, C and U, plus phosphate groups and a carbohydrate group that had not than been identified. In 1909 he isolated and identified this as a sugar group, ribose. The sugars themselves are each built around carbon rings, similar to the ones Kekulé described (see page 132), but with four carbon atoms and one oxygen atom linked, forming pentagons instead of hexagons. Levene showed that the components of the nucleic acid are themselves linked in units each made up of one phosphate, one sugar and one base, and he called these units nucleotides. Nobody knew, however, how the components of the nucleic acids were joined together.

Levene's idea was that the nucleic acid molecule was made up of a string of these nucleotides, joined together like the vertebrae of your spine, and in 1909 he gave the name ribosenucleic acid to the molecule. As there were four bases present in equal numbers in what became known as RNA, he suggested that each molecule was made up of a short chain of four nucleotides, one with each of the four bases. This became known as 'the tetranucleotide hypothesis', and held sway for decades. One consequence of this was that it reinforced the prevailing idea that the really important molecules of life are all proteins, and that the nucleic acids simply provided some sort of scaffolding to which protein molecules were attached.

It wasn't until 1929 that Levene discovered that the nucleic acid derived from thymus cells contains a different sugar group, as well as having T instead of U. Because each molecule of this sugar group has one oxygen molecule less than a corresponding ribose group, he called it deoxyribose, and the nucleic acid became known as deoxyribosenucleic acid. The names of the two nucleic acids are often shortened slightly to ribonucleic acid (RNA) and deoxyribonucleic acid (DNA). Levene still thought that the nucleotides in a DNA molecule are always linked in the same order, perhaps ACTG, ACTG, ACTG and so on. But the first clue that there is something more to DNA than mere scaffolding had already emerged in 1928, the year before DNA was named (see page 184).

EVOLUTION AT WORK

A key step towards an understanding of how evolution works was taken by Thomas Hunt Morgan and his colleagues, working at Columbia University in the second decade of the twentieth century. But it had a long pedigree.

Back in the 1870s, researchers had observed that during reproduction the nucleus of an egg cell and the nucleus of a sperm cell fuse to make a single new nucleus combining material from both parents. In 1879 the German Walther Flemming found that the nucleus contains threads of material which absorb the coloured dyes used by microscopists, so he called them chromosomes.

Each cell contains two sets of chromosomes, and in everyday cell division both sets are copied before the cell divides to make two new cells. But when egg or sperm cells are made, first there is a stage where material from the two sets of chromosomes gets mixed up, chopping bits out and joining them together in new combinations, then only one set of the 'new' chromosomes goes in to each sex cell. It is only when the egg and sperm come together to make a new cell that the full complement of chromosomes is restored, with one set from each parent. August Weismann, studying this behaviour at the University of Freiburg in the 1880s, concluded that 'heredity is brought about by the transmission from one generation to another of a substance with a definite chemical and, above all, molecular constitution', which is found in chromosomes. This is what Morgan set out to investigate.

Thomas Hunt Morgan
(1866–1945).

Morgan was doing the same kinds of experiment as Gregor Mendel (see page 134), but working with the fruit fly *Drosophila* rather than with peas. Mendel had to wait a year between generations in his studies, but the flies produce a new generation every two weeks, with the females laying hundreds of eggs at a time. The sex of the offspring is determined by one of the chromosomes – as it happens, the most easily identified. The chromosomes that determine sex come in two varieties, known as X and Y, from their shapes. In most species, the cells of females always carry the XX pair, while the cells of males carry the XY pair. So offspring always inherit one X from the mother, and either X or Y from the father. If it inherits another X it will be female, if it inherits Y it will be male. But as Morgan discovered, this is not all that those chromosomes do.

Morgan started out with a population of flies that all had red eyes, like their 'wild' counterparts.

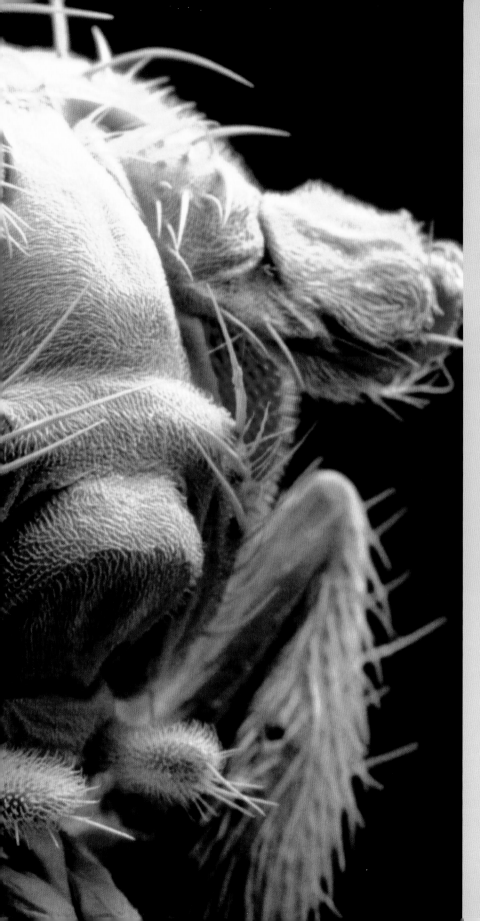

Coloured Scanning Electron Micrograph (SEM) of the head of a mutant fruit fly, *Drosophila melanogaster*.

But as a result of a chance mutation, in 1910 a single white-eyed male showed up among the thousands of flies being studied. Intrigued, Morgan mated the white-eyed male with a normal red-eyed female. All the offspring had red eyes. Then, he studied the grandchildren and succeeding generations, the way Mendel had studied peas. In the second generation, there were red-eyed females, red-eyed males, and white-eyed males, but no white-eyed females. In 1911 he concluded, after carrying out a proper statistical analysis, that whatever it was that caused the white-eye mutation must be a factor carried on the X chromosome. In the second generation females, even if one X chromosome has the mutation this is dominated by the normal factor on the other X chromosome; but in males there is no 'other' X chromosome to do the job. Further experiments showed that other properties of fruit flies are also linked with their sex, and must also be carried on the X chromosome. Morgan picked up the term 'gene', coined by the Dane Wilhelm Johannsen, for these Mendelian 'factors', developing the image of genes strung out along the thread-like chromosomes like beads strung out along a wire.

Further work showed how the process of shuffling genes to make new combinations for sex cells occurs. Paired chromosomes are chopped up, with pieces being swapped from one chromosome to the other (termed 'crossing over') then rejoined ('recombination'). Genes that are further apart along the chromosome are more likely to get separated when this process of crossing over and recombination occurs, while genes that are close together on the chromosome seldom get separated. This provided scope for mapping out the order of genes along the chromosomes, although this involved a great deal of painstaking work. It is widely accepted, though, that the key moment when the idea of Mendelian heredity and genetics became established was when Morgan and his colleagues published a classic book, *The Mechanism of Mendelian Heredity*, in 1915. Morgan continued his work on heredity, writing *The Theory of the Gene*, published in 1926, and receiving the Nobel Prize in 1933 'for his discoveries concerning the role played by the chromosome in heredity'.

Nº 66   SOMETHING TO BRAG ABOUT

At the same time that Thomas Hunt Morgan was getting to grips with the role of genes in heredity, experimenters in a seemingly unrelated field were developing the techniques that would eventually unveil the molecular mechanism of heredity. This is also a story of how a new scientific discovery can be quickly turned to experimental use, opening the way for more discoveries.

X-rays had been discovered in 1895 (see page 150), but at first their nature was something of a mystery. In 1912, however, a team headed by Max

von Laue, working at the University of Munich, found that X-rays could be diffracted by crystals, making interference patterns like the patterns made by light diffracted through the double-slit experiment (see page 79). In a key experiment, a beam of X-rays was shone through a crystal of copper sulphate and the diffraction pattern was recorded photographically. The photograph showed many distinct spots arranged in a pattern of intersecting circles, centred on the spot produced by the main beam. This helped to establish that X-rays are a form of electromagnetic radiation, like light but with shorter wavelengths, and von Laue received the Nobel Prize in 1914 'for his discovery of the diffraction of X-rays by crystals'. At the time, it was still necessary to establish the wave nature of X-rays, because many people, including a certain William Bragg, still preferred, prior to 1912, the explanation of X-rays in terms of a stream of particles, rather like cathode rays (electrons). But like any good physicist, Bragg knew that 'if it disagrees with experiment, then it is wrong'.

When news of the German experiments reached England in 1912, William Henry Bragg was an established physicist working at Leeds University. His son, William Lawrence Bragg (always known as Lawrence) was just starting out as a research physicist in Cambridge. Intrigued by von Laue's work, and open to new

William Henry Bragg (1862–1942).

ideas, they discussed the implications with one another, and realized that it ought to be possible to work out the structure of a crystal by analysing the pattern of bright and dark spots produced by diffraction. They decided to divide up the task. Lawrence worked out the rules which made it possible to predict exactly where the bright and dark spots would be located if a beam of X-rays with a particular wavelength struck a crystal made up of atoms spaced a certain distance apart from one another. This became known as Bragg's law. It meant that if you know the way atoms are spaced out in a crystal you can use diffraction to measure the wavelength of the X-rays, or if you know the wavelength of the X-rays you can use diffraction to work out how the atoms in a crystal are arranged. He was immediately able to use his law to make a partial interpretation of the diffraction patterns obtained in Munich, but proper analysis required more information about the wavelengths of the X-rays. So William concentrated on the experimental side, including inventing the first X-ray spectrometer, an instrument to measure the wavelengths to be plugged in to Bragg's Law.

Interpreting the data is horribly difficult for complicated structures in which there are large numbers of different kinds of atoms. But simpler crystals are much easier to understand, and this kind of work showed, for example, that crystals of sodium chloride (common salt, NaCl) are made up not of lots of distinct NaCl molecules but of an array of sodium (Na) and Chlorine (Cl) with equal spacing, alternating in a lattice. The work of the two Braggs, which established that X-rays could be used as a tool to investigate the structure of matter, culminated in a book, *X-rays and Crystal Structure*, published in 1915, while Lawrence was serving in the British Army in France. It was there that he learned that had shared the Nobel Prize for 1915 with his

LEFT **A model of the atomic structure of crystalline common salt (sodium chloride).**

RIGHT **The first X-ray crystal diffraction photograph, made in 1912 by the German physicist Max von Laue (1879–1960).**

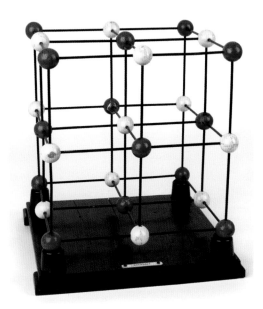

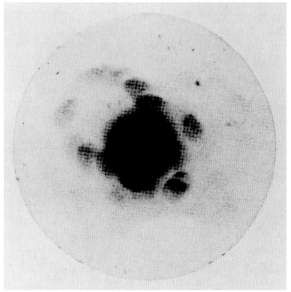

father 'for their services in the analysis of crystal structure by means of X-rays'. Lawrence Bragg is the youngest person, at 25, to receive the Nobel Prize in Physics. As he said in his Nobel Lecture: 'The examination of crystal structure, with the aid of X-rays has given us for the first time an insight into the actual arrangement of the atoms in solid bodies ... There seems to be hardly any type of matter in the condition of a true solid which we cannot attempt to analyse by means of X-rays. For the first time the exact arrangement of the atoms in solids has become known; we can see how far the atoms are apart and how they are grouped.'[34]

It was this ability that would, in the decades that followed, lead to an understanding of the structure of proteins (see page 195) and DNA itself (see page 224).

Nᵒ· 67   LIGHT FROM THE DARKNESS

T he famous test of Albert Einstein's general theory of relativity carried out during observations of a solar eclipse in1919 is a classic example of the scientific method at work. A new idea – in this case, the general theory – makes a prediction, which is tested by experiment. It passes the test, and becomes established as a good scientific theory.

Before Einstein came along, the best theory of how the Universe worked was based on Isaac Newton's laws of motion. These describe the way things move in an unchanging backdrop of space as time passes, with time ticking away at a steady rate like some great universal clock. Einstein's general theory of relativity, which he completed at the end of 1915, was also a description of how things move in the Universe, but in his theory the rate at which clocks tick depends on how fast they are moving, and space itself is curved by the presence of matter.

Among other things, this means that light from a distant star passing close by the surface of the Sun would be deflected by a tiny angle, because of the way the mass of the Sun bends space. Einstein was well aware of this, and used his theory to calculate the size of this deflection. As it happens, Newton's theory also predicts that light will be deflected as it passes by the Sun, because of the gravitational influence of the Sun on what Newton thought of as particles of light. But, crucially, the size of the deflection predicted by Einstein's theory is exactly twice as large as the deflection predicted by Newton's theory. The effect predicted by Einstein was a deflection of 1.75 seconds of arc; Newton's theory predicts 0.875 arc seconds. Here was a clear-cut way to see which theory provided a better description of the Universe, if only there were a way to look at stars almost directly 'behind' our Sun. The way to do this, of course, is to

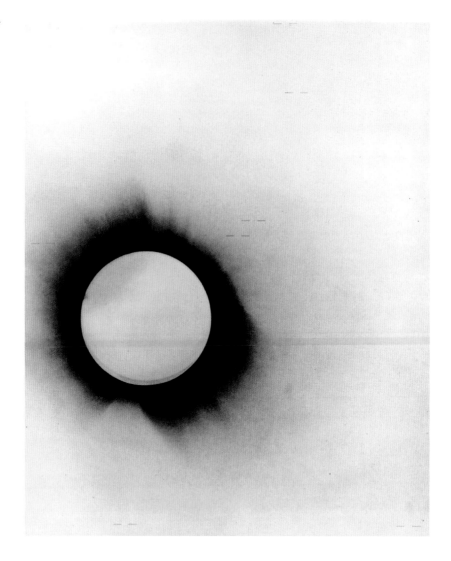

Photograph used by Arthur Eddington (1882–1944) to measure the apparent shift in position of the marked stars, caused by the light-bending effect predicted by the general theory of relativity.

photograph the stars when the light from the Sun itself is blotted out by the Moon during a solar eclipse, and compare the prints with photographs of the same part of the sky taken when the Sun is not in the way. Einstein knew this, but with the First World War raging in Europe, he was in no position to arrange for observers to make the test.

News of Einstein's ideas travelled quickly to scientists in neutral Holland, and from them to astronomers in England. There, Arthur Eddington realized that a suitable eclipse would be visible from the island of Principe, off the west coast of Africa, in 1919. He obtained permission to organize an expedition to test Einstein's theory, *provided* that the fighting in Europe stopped in time. When the Armistice was declared in November 1918, even though Britain was still technically at war (the peace treaty was not signed until June 1919), he was given the go ahead, and just had time to make the arrangements. He was able to obtain the crucial photographs on 29 May 1919. Even though the observations were hampered by cloud, they were just good enough for his purpose. When the photographic plates were compared with photographs of the same part of the sky when the Sun was not in the line of sight, they showed exactly the amount of deflection predicted by Einstein – twice as much as predicted by Newtonian theory. Newton's theory is still good for many things – such as calculating the fall of an apple from a tree or even the orbit of a spaceship on its way to Mars – but it is not as good as Einstein's theory at describing the Universe at large.

Eddington announced his results to a joint meeting of the Royal Society and the Royal Astronomical Society in London on 6 November 1919, just four years after Einstein had presented his general theory to scientists in Berlin. Since then, the general theory has passed many other tests, and is still regarded as the best theory of time, space and the Universe that we have. The light-bending effect itself is now seen on very large scales, where whole galaxies of stars, and even clusters, act as 'gravitational lenses' to focus the light from even more distant objects and make them visible to our telescopes.

There is also a more practical application of the general theory. The instruments used by GPS satellites to work out the positions of objects on Earth are so precise that as they move in the Earth's gravitational field they have to take account of the effects described by the general theory. Every time you use sat nav, or the 'find location' facility on a smartphone, you are using – and, in a sense, testing – the general theory of relativity.

Nᵒ· 68 ELECTRON WAVES AND QUANTUM DUALITY

During the 1920s, a combination of theory and experiment led to a dramatic advance in our understanding of the world of the very small – atoms and below – in the form of quantum mechanics. The best example is the discovery that electrons are both particle *and* wave.

Albert Einstein's explanation of the photoelectric effect (see page 158) had led to the identification of a simple equation relating the wavelength of a particular colour of light to the momentum of the particle of light (photon) with that colour. Wavelength is, of course, a property of waves; momentum is a property of particles. Light seemed to be both wave and particle, but at first this looked like a peculiarity of light alone. Then, in 1924 (in his PhD thesis) the French physicist Louis de Broglie put forward the idea that Einstein's equation could be turned around, to give a wavelength of any particle in terms of its momentum. In particular, this implied that electrons should behave like waves under certain

Clinton Joseph Davisson
(1881–1958).

circumstances. The key equation is $\lambda = h/p$, where λ is wavelength, p is momentum, and h is a very small number known as Planck's constant. Because h is so small, the 'wave-particle duality' is predicted to show up for only tiny objects, the size of atoms or smaller.

Even before de Broglie published his idea, Clinton Davisson, an experimenter at the Bell Laboratories in America, had been studying the way electrons bounced off (or scattered from) a nickel surface. On a visit to Britain in 1926, he was astonished to hear the German theorist Max Born give a lecture in which he quoted some of Davisson's published results as supporting de Broglie's idea. Back in the USA, late in 1926 he set out with his assistant Lester Germer to make a proper test, using essentially the same techniques that the Braggs had used to study X-rays (see page 177). In the American experiment, which produced results early in 1927, electrons from a hot wire were accelerated by electric fields and fired in a beam at a nickel surface. This 'target' could be rotated to different angles, and the electron detector used to monitor the scattered electrons could also be moved to different places to observe the electrons coming back from the surface. They found distinct peaks in the intensity of the scattered electron beam at certain angles, exactly matching Bragg's Law for waves diffracted by the atoms spaced out in the crystal lattice. The wavelength implied was exactly the wavelength obtained from de Broglie's calculation.

Meanwhile, the British physicist George Thomson (the son of J. J. Thomson, see page 152), based in Aberdeen, had also heard Born's lecture, and set out to search for evidence of electron waves by firing electrons through thin pieces of gold foil (between one ten-thousandth and one hundred-thousandth of a millimetre thick) in a cathode ray tube. Early in 1927 he produced a diffraction pattern on photographic plates, and proved that it was caused by electrons, not X-rays, by bending the beam with electric and magnetic fields. It agreed with the predictions from de Broglie's calculations to within 1 per cent. Two subtly different experiments had almost simultaneously proved that de Broglie was right. Or, as Davisson put it in his Nobel Lecture: 'That streams of electrons possess the properties of beams of waves was discovered early in 1927 in a large industrial laboratory in the midst of a great city, and in a small university laboratory overlooking a cold and desolate sea ... Discoveries in physics are made when the time for making them is ripe, and not before.'[35]

Davisson and Thomson shared the Nobel Prize in Physics in 1937 'for their experimental discovery of the diffraction of electrons by crystals'. De Broglie had already received the prize, in 1929, 'for his discovery of the wave nature of electrons'. But physicists were left struggling to understand how subatomic entities could be both particle and wave. In the course of this struggle, they developed two versions of quantum mechanics. One, essentially based on the particle idea, was developed by Werner Heisenberg and his colleagues; the other version, based on the wave idea, was developed by Erwin Schrödinger. The two theories are mathematically equivalent, and they are both good theories because they each give the 'right' answers to calculations involving the behaviour of things like electrons when compared with experiments. But it remains a mystery how the world can be like that. J. J. Thomson was awarded the Nobel Prize for proving electrons are particles; his son was awarded the Nobel Prize for proving electrons are waves. Both were right. Don't worry if you cannot see how this is possible. As another Nobel Prize-winner, Richard Feynman, used to say, 'Nobody understands quantum mechanics.'

Electron diffraction pattern produced by beryllium.

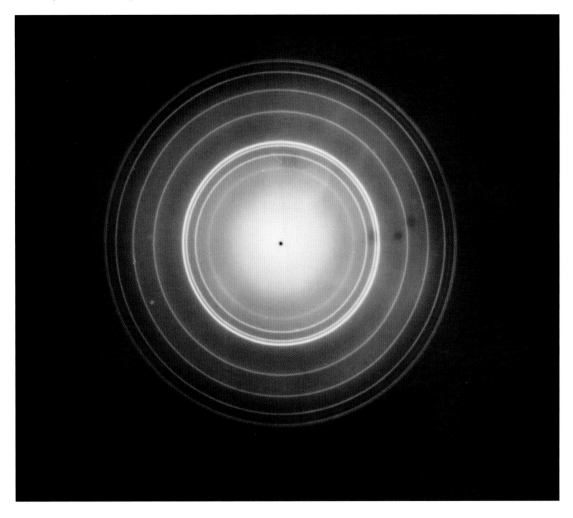

TAKING THE ROUGH WITH
THE SMOOTH

The experimental line that would lead to an understanding of heredity and
genetics started with Gregor Mendel's work with peas (see page 134) and
continued with Thomas Hunt Morgan's studies of fruit flies (see page 173). The
next development, taken in 1928 by Frederick Griffith, a medical officer working
for the UK Ministry of Health in London, brought biologists a step closer to the
key molecules involved. Mendel's peas produced only one generation a year,
limiting his opportunity to study heredity. Morgan's fruit flies reproduced every
couple of weeks. But bacteriologists can see changes that take place in a matter
of hours. Griffith was interested in bacteria as agents of disease, rather than as a
tool for research into genetics. But that did not stop him making a discovery that
turned out to be crucial in the study of genetics.

Following the global influenza epidemic of 1918–1920, which killed at least
50 million people (more than the combined battlefield casualties of the First World
War), governments around the world stepped up their research into the causes and
cures of infectious diseases. During the 1920s, Griffith had been investigating the
possibility of developing a vaccine against pneumonia. He was working with two
strains of pneumococcus bacteria, and studying how they affected mice. In one
strain the bacteria are covered in a smooth coating (a polysaccharide), which
makes cultures of the strain look shiny. As a result this strain is prosaically called
'smooth' (S). The other strain used by Griffith has a rough surface, and is called

Influenza medical advice
poster, Washington DC,
USA, 1918. The 1918 Spanish
flu pandemic infected
one fifth of the world's
population and killed around
50 million people.

'rough' (R). Cultures of this strain have a lumpy
appearance. (There is also a third strain of pneumonia-
causing bacteria – pneumococci – but these were not used
in his experiments.) Before Griffith's work, bacteriologists
thought that each of the three strains of pneumococci
were completely independent of each other, each fixed
with its own properties down the generations. But Griffith
knew that different stains of pneumococci, some lethal and
some not, could be present at the same time in the body of
a person (or mouse) with pneumonia, and considered the
possibility that instead of different varieties being present
from the beginning, one kind might change into another.

When mice (or, indeed, people) are infected with the
rough strain of pneumococci, the invaders are easily
recognized by the body's immune system, which kills
them off before any serious harm is done. But the covering
on the smooth strain seems to disguise them from the
immune system, with the result that they can proliferate

TREASURY DEPARTMENT
UNITED STATE'S PUBLIC HEALTH SERVICE

INFLUENZA

Spread by Droplets sprayed from Nose and Throat

Cover each COUGH and SNEEZE with hand-
kerchief.

Spread by contact.

AVOID CROWDS.

If possible, WALK TO WORK.

Do not spit on floor or sidewalk.

Do not use common drinking cups and common
towels.

Avoid excessive fatigue.

If taken ill, go to bed and send for a doctor.

The above applies also to colds, bronchitis,
pneumonia, and tuberculosis.

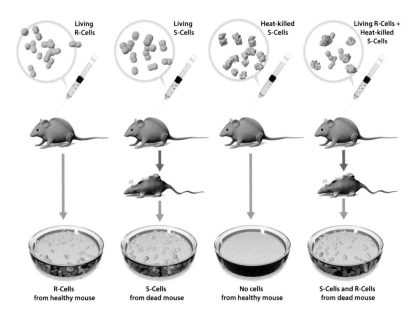

Illustration of the experiment reported in 1928 by British bacteriologist Frederick Griffith (1879–1941), providing evidence that bacteria can transfer genetic information through a process called transformation.

and cause serious illness, even death. In a series of experiments, Griffith showed that mice injected with the rough strain of pneumococci lived, mice injected with the smooth strain died, and that mice injected with bacteria from the smooth strain that had been killed by heating also lived. But then came the bombshell that he reported in January 1928.

When Griffith mixed heat-treated smooth bacteria with live rough bacteria and injected them into mice, the mice died. On their own, neither of the two forms in the mixture was a killer, but together they were lethal. Samples taken from the dead mice showed that they were teeming with live smooth pneumococci. Somehow, the live rough bacteria had been 'transformed', in Griffith's word, into live smooth bacteria. The explanation he proposed was that what we would now call genetic material from the dead smooth bacteria had been passed into the living rough bacteria, enabling them to 'learn' how to develop a smooth coating. Further experiments confirmed that this must be the case, and Griffith called the substance involved the 'transforming factor'. Once they had been transformed in this way, when they were transferred to a dish in the laboratory and monitored closely, the 'new' smooth bacteria were seen to breed true, replicating to produce a colony of smooth bacteria. As Griffith wrote in the scientific paper announcing the discovery, 'The R form … has been transformed into the S form'. He realized that the transforming factor must be a chemical substance that was not affected by the heat treatment. But he did not know which molecules were involved in the genetic transfer. That only became clear after 1944 (see page 213), as a result of new experiments directly inspired by Griffith's observations. But he did not live to see these developments; he was killed in an air raid during the London Blitz of 1941.

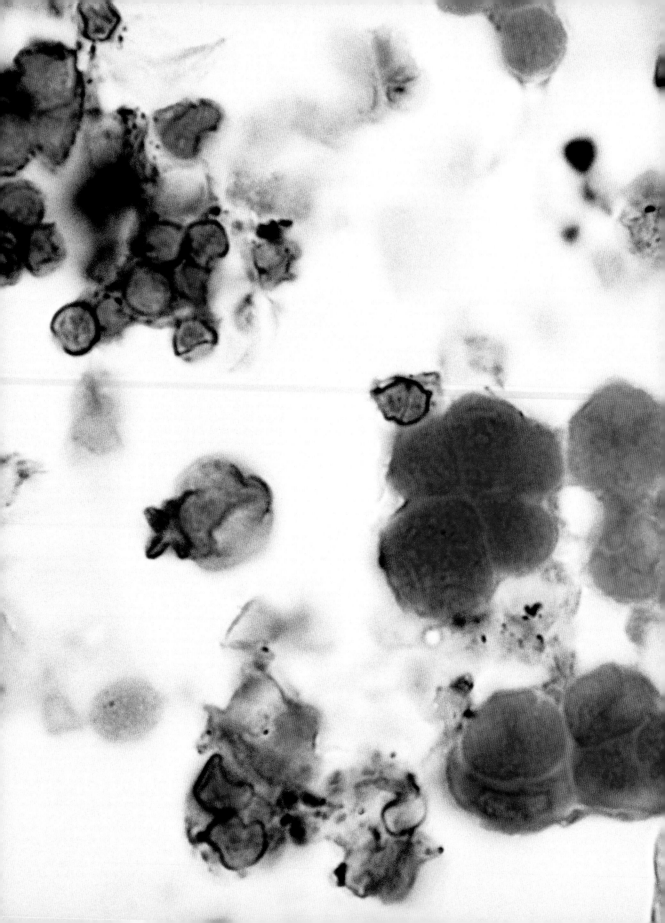

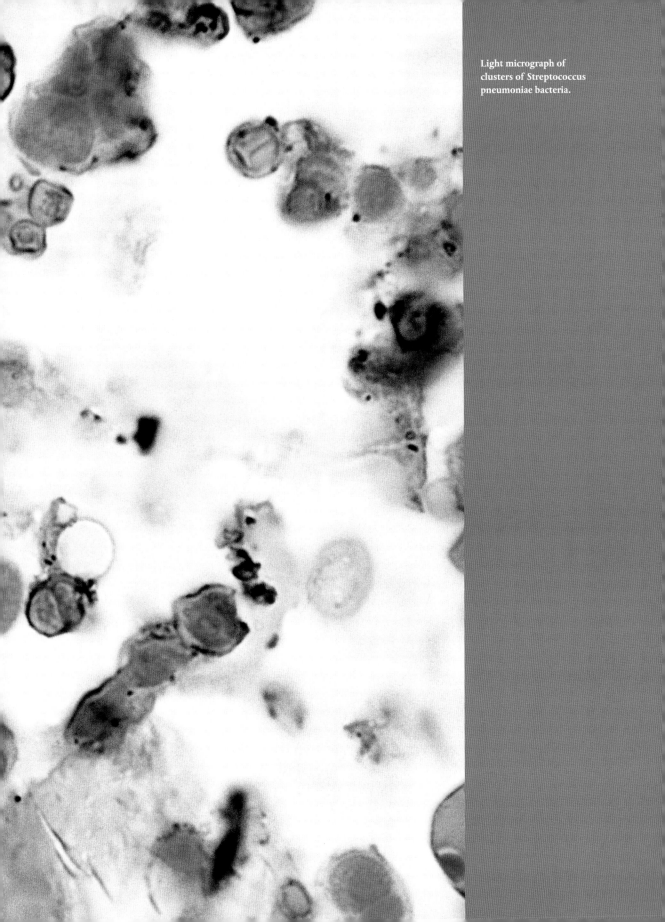

Light micrograph of clusters of Streptococcus pneumoniae bacteria.

AN ANTIBIOTIC BREAKTHROUGH

The discovery of the antibiotic properties of penicillin and the application of the discovery is one of the most confusing, as well as one of the most important, in the history of biochemistry. But if any one 'experiment' can be pinpointed as the moment of discovery, it was when Alexander Fleming noticed something odd in a dirty Petri dish in his laboratory at St Mary's Hospital, in London, in 1928.

Fleming already had something of a track record of serendipitous discovery. Like many researchers in the years following the First World War (see page 184), in the early 1920s Fleming was searching for any chemical agent with bacteria-killing properties. The cultures that they studied were grown in Petri dishes – shallow, circular glass dishes. One day in 1922, Fleming had a runny nose, which dripped mucus onto one of these plates. The nasal fluid killed the bacteria, leading him to discover lysozyme, an enzyme present in tears, and which is a natural antibacterial agent. Unfortunately, though, the microbes that are most strongly affected by lysozyme do not infect people.

In September 1928, Fleming returned from a holiday and set about clearing up a jumble of Petri dishes that he had left behind uncleaned. One of them, which had been seeded with staphylococci, contained a spot of mould, with a clear circle around the mould, where the staphylococci had been killed. Fleming grew the mould in culture, identifying it as a member of the genus Penicillium, and finding that it killed a variety of bacteria. It later turned out that a colleague in a lab on the floor below had been working with Penicillium, and it seems that a spore from his lab had blown out of the window and in through the window of Fleming's lab, both of which were open in the summer heat.

Fleming made an antibacterial broth from the mould, which he initially called 'mould juice', before naming it 'penicillin' on 7 March 1929. But although he published news of his discovery and carried out some further experiments, Fleming did not pursue the investigation of penicillin, partly because it turned out to be ineffective against typhoid and paratyphoid, the diseases he was working on at the time. It was also very difficult to produce penicillin in quantity. As Fleming later said: 'When I had some active penicillin I had great difficulty in finding a suitable patient for its trial, and owing to its instability there was generally no supply of penicillin if a suitable case turned up. A few tentative trials gave favourable results but nothing miraculous and I was convinced that before it could be used extensively it would have to be concentrated and some of the crude culture fluid removed.'[36]

Nevertheless, when a junior colleague of Fleming, Cecil George Paine, moved from St Mary's to Sheffield in 1929 he took knowledge of the discovery with him. In 1930, he used penicillin to clear up infections in babies at the

Sheffield Royal Infirmary, but he did not publish news of this success, and with the same difficulties in producing penicillin in quantity he did not follow up the achievement. His notes were only rediscovered in 1983.

Although Fleming did attempt to interest chemists in developing techniques to produce large quantities of penicillin, no progress was made until 1938, when a team of researchers at the Dunn School of Pathology, in Oxford, had completed a project to crystallize lysozyme. They decided to investigate other natural antibacterial agents, in the mistaken belief that they must all be enzymes, like lysozyme. One of the agents they chose to study was penicillin. They developed a technique for concentrating penicillin, and published their results in *The Lancet* in August 1940. By then, with the Second World War raging, the immediate importance of the work was recognized, and funding was provided for the research into mass production of the medicine to be stepped up, initially in Britain and then in the United states. By 1944, 2.3 million doses of penicillin were available in time for the D-Day invasion of Normandy. The two key researchers involved at the Dunn School were Howard Florey and Ernst Chain, who shared the Nobel Prize with Fleming in 1945 'for the discovery of penicillin and its curative effect in various infectious diseases'.

In a sobering footnote to his Nobel Lecture, Fleming gave a warning which resonates in these days of drug-resistant bacteria: 'Mr. X. has a sore throat. He buys some penicillin and gives himself, not enough to kill the streptococci, but

LEFT Photograph of the original culture plate of the fungus *Penicillium notatum*, made by the Scottish bacteriologist Alexander Fleming (1881–1955).

RIGHT A drawing of the original culture plate of the fungus *Penicillium notatum*, made by Fleming, with his notes.

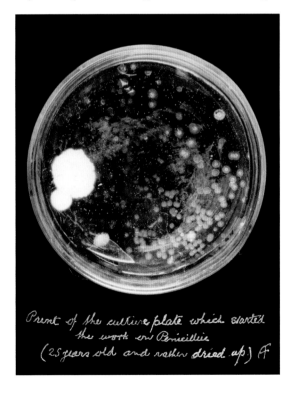

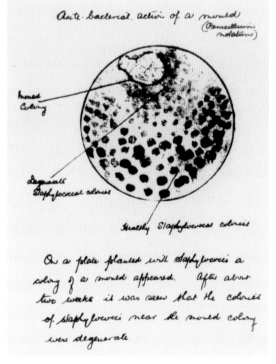

enough to educate them to resist penicillin. He then infects his wife. Mrs. X gets pneumonia and is treated with penicillin. As the streptococci are now resistant to penicillin the treatment fails. Mrs. X dies. Who is primarily responsible for Mrs. X's death? Why Mr. X, whose negligent use of penicillin changed the nature of the microbe.'[37]

Nº. 71   SPLITTING THE ATOM

The important thing about so-called 'atom-splitting' experiments is that they actually split the nucleus, the tiny central kernel of the atom first identified by Ernest Rutherford (see page 165). It is the number of positively charged protons in the nucleus that determines the nature of the atom – what element it belongs to. Hydrogen has one proton per atom, helium two, and so on. The electric charge of these protons is balanced by the negative electric charge of an equal number of electrons in a cloud in the outer part of the atom – one for hydrogen, two for helium, and so on. The arrangement of electrons in the cloud determines the chemical properties of an element. The nucleus also contains neutral particles called neutrons, which do not affect the chemistry. Atoms with different numbers of neutrons but the same number of protons are called isotopes of a particular element. For example, some atoms of helium have two protons and one neutron, and are known as helium3; others have two protons and two neutrons and are known as helium4.

The particles in the nucleus are held together by an interaction known as the strong nuclear force, which has only a very short range but which affects both neutrons and protons and is powerful enough at short range to overcome the natural tendency of the positively charged protons to repel one another. But the bigger a nucleus is (the more protons it contains), the harder it is for the strong force to overcome the electric repulsion, which makes it possible for heavy nuclei to split, or fission, into two or more lighter pieces. This is the process that people usually think of today when talking about splitting the atom – the kind of fission involved in a nuclear bomb or a nuclear power station (see page 204).

Ernest Walton (1903 – 1995) in the lead-lined box where he sat to study 'scintillations' produced by alpha particles from nuclear fission.

But it is also possible to force nuclei to split by bombarding them with high-energy particles from outside. This is what John Cockcroft and Ernest Walton did in the first atom-splitting experiments, at the Cavendish Laboratory in Cambridge on 14 April 1932. Rutherford was the Director of the laboratory at the time, and had suggested that Cockcroft and Walton should combine their efforts on the project.

The realization that such a project might be feasible was inspired by a visit to Cambridge by the Russian theorist George Gamow in 1929. Gamow had calculated that under the right circumstances a relatively low energy particle could penetrate an atomic nucleus and trigger interactions there. Cockcroft worked out that a proton accelerator running at a few hundred thousand volts might do the job, and it seemed that such an accelerator might be built with the limited resources available at the Cavendish. As Cockcroft later said, in a speech at a Nobel banquet, new ideas combined with new technology to make many great discoveries possible at that time.

Cockcroft and Walton brought separate skills to the task. Cockcroft was a theorist who worked out what was possible, while Walton was a superb experimenter, a 'handson' man, who built an apparatus that accelerated protons (hydrogen nuclei) down a tube across an electric potential of just over 700 kilovolts. This was an early example of a so-called 'linear accelerator'. The proton beam was directed at a target made of lithium, a soft metal in which each nucleus has three protons and either three or four neutrons. The bombardment produced traces of helium, atoms with only two protons in each nucleus. The helium was in the form of fast-moving alpha particles (essentially helium nuclei), which Walton detected by watching the flashes made by these particles when they hit a paper screen coated with zinc sulphide.

To do this, he sat in a lead-lined box, to protect him from radiation, peering through a microscope focused on the screen. In order to work out how much lead was needed, a zinc sulphide screen was hung on the wall inside. If it glowed, another layer of lead was added! Walton wrote: 'In the microscope there was a wonderful sight. Lots of scintillations, looking just like stars flashing out momentarily on a clear dark night.' Rutherford himself had investigated and named alpha radiation at the end of the nineteenth century. When he was brought in to see the experiment, he said: 'Those look mighty like alpha particles to me.' A crucial clue was that the flashes occurred in pairs, showing that two alpha particles had been emitted simultaneously from the same source. This became known as 'splitting' the atom, but it was actually a two-stage process in which an incoming proton briefly combines with a lithium nucleus to make a nucleus of beryllium, containing four protons, then the beryllium nucleus splits into two helium nuclei. The team shared the Nobel Prize in 1951, 'for their pioneer work on the transmutation of atomic nuclei by artificially accelerated atomic particles'.

MAKING VITAMIN C

A lthough the importance of what became known as vitamin C had been known for many years (see page 42), it was only in the 1930s that the chemical structure of this substance, ascorbic acid, was worked out. This enabled the team that had determined the structure to develop a way to synthesize the vitamin – the first vitamin to be manufactured artificially.

The leader of the team was Norman Haworth, based at the University of Birmingham in England. In the late 1920s, Haworth was investigating the structure of carbohydrates. These are compounds built up from carbon and water (hence the name). The simplest carbohydrates are sugars, but the basic sugar units can be linked to make more complex molecules including things like starch, and the cellulose that forms the structure of plants. Each molecule of the kind of sugar found in grapes contains just 6 atoms of carbon, 6 of oxygen and 12 of hydrogen, but a single molecule of cellulose contains thousands of atoms. It was only in 1925 that the structures of the most basic sugars were determined (they are built around rings, like benzene; see page 132), opening the way for Haworth and others to work out the structure of more complex carbohydrates.

Microscopic view of Vitamin C crystal.

**Tadeus Reichstein
(1897–1996).**

At the end of the 1920s, Albert Szent-Györgyi, a Hungarian initially working at Groningen, in the Netherlands, and afterwards based in Cambridge, isolated a compound which he called hexuronic acid from animal adrenal glands and the juice of plants including oranges and cabbages. He had been studying the function of the adrenal system and the way in which a failure of the system causes the fatal Addison's disease. Victims become marked with a brown pigmentation which reminded him of the way some fruits, such as apples and bananas, turn brown when they decay. This brown colour is related to oxidation, so he decided to study plants that do not turn brown in this way, to find out what stops the oxidation. He found that these plants contain a powerful reducing agent which stops the development of the brown material. As he later remarked, 'There was great excitement in my little basement room in Groningen, when I found that the adrenal cortex contained a similar reducing substance in relatively large quantities.'[38] This was the hexuronic acid. Szent-Györgyi suspected that he had found vitamin C, but he lacked the facilities to test this speculation.

Szent-Györgyi went back to Hungary at the beginning of the 1930s, where he discovered that he could also obtain hexuronic acid from paprika, a common ingredient of Hungarian cooking. Experiments there showed unmistakably that hexuronic acid prevents scurvy (that is, it is an antiscorbutic). In April 1932 his team reported that they had been able to protect guinea pigs from scurvy by dosing them daily with one milligram of hexuronic acid. Also, the antiscorbutic activity of plant juices corresponded to the amount of hexuronic acid they contained. So, once Haworth had determined its structure, the name was changed to ascorbic acid.

Haworth was an obvious person to analyse the acid, because it has chemical similarities to acids derived from sugars. The basic chemical formula of ascorbic acid is $C_6H_8O_6$, but the question was how those atoms are arranged in three dimensions. Szent-Györgyi visited Haworth to discuss his discoveries, and provided him with a sample of hexuronic acid. The Birmingham team tackled the puzzle from several directions.

Key evidence came from the still relatively new science of X-ray crystallography (see page 176), which showed that the molecules have an unusually flat, almost two-dimensional structure. By analysing the products produced when ascorbic acid reacted with other substances (for example, by oxidation), the Birmingham team were able to show that the molecule is made up of a short chain of carbon atoms attached to a five-sided ring, built around four carbon atoms and one oxygen atom, with other atoms attached at the corners. The oxidation process they used involved ozone, the form of oxygen in which there are three atoms per molecule, rather than the more common di-atomic form of oxygen.

In 1933, the year after he had identified the structure of vitamin C, Haworth was able to synthesize it from chemical raw materials. Another British chemist,

Edmund Hirst, also synthesized the vitamin; but the key technique was developed by the Pole Tadeus Reichstein. In 1934 the rights to the Reichstein process were bought by the firm Hoffman-La Roche, who began marketing vitamin C under the name 'Redoxon'.

Szent-Györgyi received the Nobel Prize in Physiology or Medicine in 1937, 'for his discoveries in connection with the biological combustion processes, with special reference to vitamin C and the catalysis of fumaric acid'. The same year, Haworth received the Nobel Prize in Chemistry 'for his investigations on carbohydrates and vitamin C'.

Nᵒ. 73 PROBING PROTEINS

The way improvements in technology lead to advances in science is highlighted by the way X-ray crystallography (see page 238) was applied to the study of biological molecules. The first biomolecules to be investigated in this way were the proteins, complex molecules which were thought, in the 1930s, to be the key carriers of biological information, as well as providing structural material for the body. Proteins are long chains made up of sub-units called amino acids. All proteins in all living things on Earth are made up of different combinations of just a score of these units, arranged in different ways. Some idea of their complexity is provided by the fact that while weights of amino acids are typically about 100 units on a scale where an atom of hydrogen has one mass unit, molecular weights of proteins range from a few thousand to several million units. It was crystallography that would eventually reveal that the key feature of these long chain molecules is that they fold up into complex shapes, and these shapes determine their biological properties.

The first steps towards this understanding were taken by J. D. Bernal and his colleagues in Cambridge in 1934. Bernal had worked with William Bragg in the 1920s, and cut his scientific teeth by using X-ray crystallography to determine the structures of graphite and bronze. At the Cavendish laboratory, he turned his attention to organic molecules, but ran into a problem with proteins because when these were dried out ready for study their structure collapsed, like a structured house of cards collapsing into a disordered heap.

The standard way of preparing crystals is to grow them in a concentrated solution in water, known as the 'mother liquor'. To crystallize a protein, the purified protein is allowed to settle out of such a concentrated solution. The individual protein molecules align themselves in a repeating series of 'unit cells' with a regular pattern, forming a crystalline 'lattice'. John Philpott, a researcher based in Uppsala, in Sweden, was doing just this with the protein pepsin in the mid-1930s (pepsin is a digestive enzyme that breaks down other proteins in our

food). When he went on a skiing holiday he left some crystals in their mother liquor in the refrigerator in his lab, and when he got back was surprised to find how much they had grown. A visitor from Cambridge, Glen Millikan, took some of these crystals, still in a tube containing the mother liquor, back to the Cavendish with him in his coat pocket, and gave them to Bernal.

Working with Dorothy Crowfoot (who later married and became known as Dorothy Crowfoot Hodgkin), Bernal first studied a crystal exposed to air while being irradiated, but found no diffraction. He had, however, also been monitoring the crystal through a microscope using polarized light. When the crystal was fresh and damp, this showed a feature known as birefringence, which indicates an ordered crystal structure, but the birefringence disappeared as the crystal dried out. So they carried out more experiments with both the crystal and its mother liquor sealed inside a thin-walled glass capillary tube. This provided the first X-ray diffraction photograph of single pepsin crystals, still in their mother liquor, in 1934. Bernal's sealed capillary technique became the standard way to collect diffraction data on large molecules of this kind for the next fifty years.

Bernal realized that photographs like this might be interpreted to reveal the structure of the protein molecules themselves. As Bernal and Crowfoot said when describing their experiment in the journal, *Nature*: 'Now that a crystalline protein has been made to give X-ray photographs, it is clear that we have the means of checking them and, by examining the structure of all crystalline proteins, arriving at far more detailed conclusions about protein structure than previous physical or chemical methods have been able to give.'[39]

Dorothy Crowfoot Hodgkin went on to play a major role in the development of the application of X-ray diffraction crystallography to the study of biologically important molecules over the next few decades (see page 238). But this was an immensely difficult task, for reasons that slowly became clear in the years that followed. What is called the primary structure of a protein is the order of the amino acids along a chain. These chains can then be twisted around to make, for example, a helix, which is the secondary structure. And then the helix is twisted into a kind of knot in three dimensions, the tertiary structure. By 1971, only seven proteins had had their structure fully worked out. Since then, improving technology, including powerful computers, has led to the detailed structural analysis of more than 30,000 proteins.

№ 74   ARTIFICIAL RADIOACTIVITY

Marie Curie had investigated and named radioactivity at the end of the nineteenth century (see page 154). In the 1930s, her daughter, Irène, discovered a way to manufacture radioactive elements artificially. Irène worked with her husband, Frédéric. He had originally been known as Frédéric Joliot, but when they married Irène and Frédéric each took the surname Joliot-Curie.

The Joliot-Curies worked at the Radium Institute in Paris, which had been founded by Irène's mother, and is now known as the Curie Institute. There, they had access to the world's largest supply of the highly radioactive element polonium, discovered and named by Irène's parents. Polonium is a strong emitter of alpha particles, and in the 1930s the Joliot-Curies were investigating the effect of this radiation on other elements. In January 1932, their experiments showed that when beryllium was bombarded with alpha particles it emitted another kind of radiation, which was very difficult to detect but which could knock protons (which are easy to detect) out of the nuclei of the atoms in paraffin. James Chadwick, working at the Cavendish Laboratory, followed this

Frederic Joliot-Curie (1900–1958) and Irene Joliot-Curie (1897–1956).

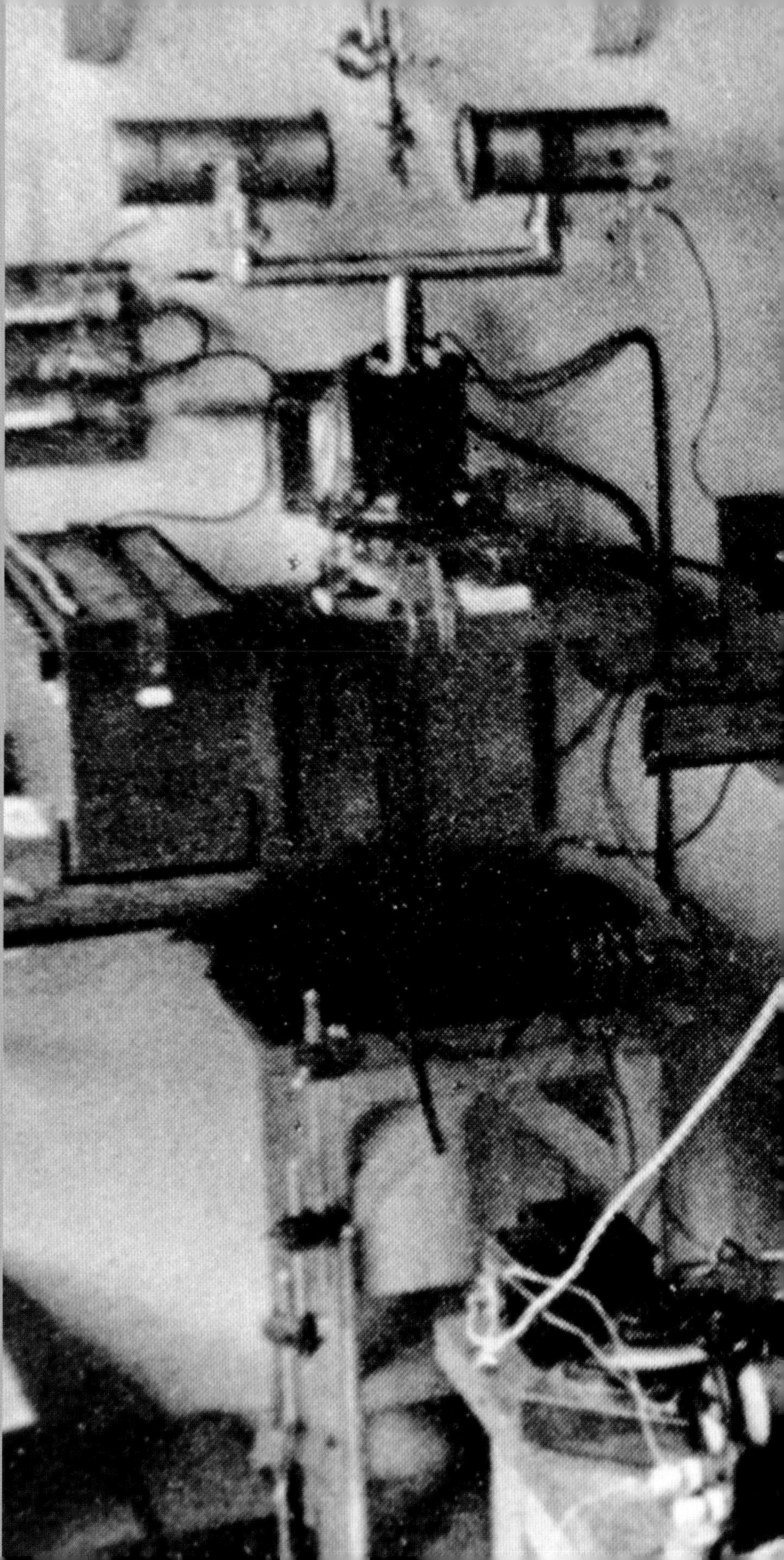

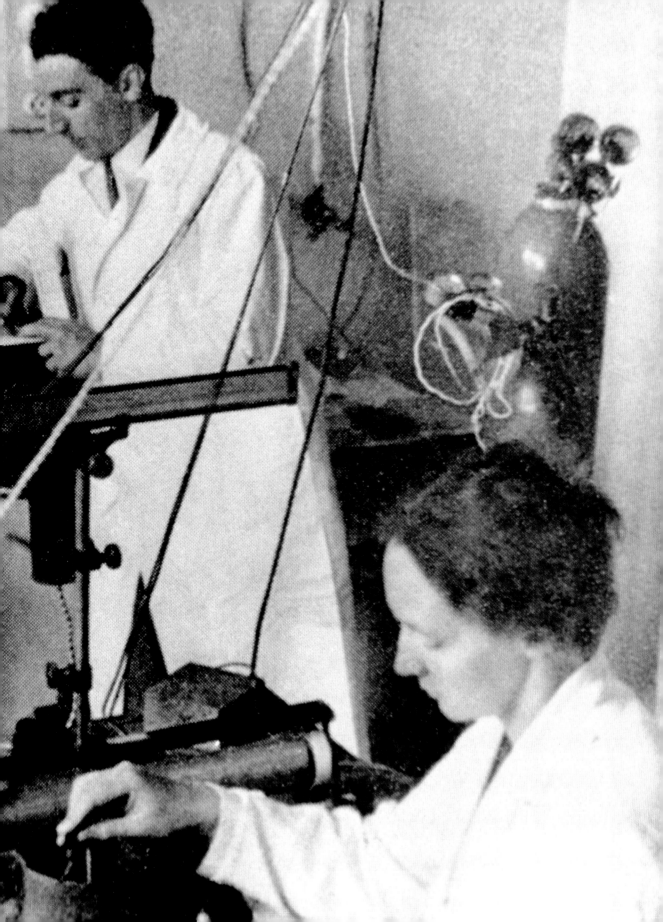

up and within a few weeks had identified the previously unknown radiation as a stream of electrically neutral particles, each with about the same mass as a proton, which became known as neutrons. In 1935 he received the Nobel Prize in Physics for the discovery.

Meanwhile, Irène and Frédéric had continued their experiments. At the beginning of 1934, while studying the effect of alpha radiation on aluminium, they found that positively charged particles (now known as positrons, a positively charged counterpart to negatively charged electrons) continued to be emitted even after the alpha radiation was stopped. The number of positrons emitted decreased by half every three minutes.

This was an important clue to what was going on. By then, it had been well established (chiefly by Ernest Rutherford) that any naturally occurring radioactive element 'decays' in this way. However much, or however little, of the element you start with, half the nuclei transform into nuclei of another element and spit out radiation (such as positrons) in a certain time; half of the remainder decay in the next interval the same length, and so on. Each radioactive element has a characteristic 'half life', some are measured in fractions of a second, some are measured in millions of years. So the Joliot-Curies knew that they were observing true radioactivity from an element with a radioactive half life of about three minutes. Their experiment involved both the transmutation of one element into another, and the artificial creation of radioactivity. As Irène later explained: 'Returning to our hypothesis concerning the transformation of the aluminium nucleus into a silicon nucleus, we have supposed that the phenomenon takes place in two stages: first there is the capture of the alpha particle and the instantaneous expulsion of the neutron, with the formation of a radioactive atom which is an isotope of phosphorus of atomic weight 30, while the stable phosphorus atom has an atomic weight of 31. Next, this unstable atom, this new radio-element which we have called 'radio-phosphorus' decomposes exponentially with a half-life of three minutes.'[40]

Before this discovery, radioactivity was a property associated only with about thirty naturally occurring elements. The artificial creation of radioactive elements opened up a whole new field of research. The Joliot-Curies quickly extended their work to experiments involving other elements, and found that they could produce radioactive elements from boron and magnesium; bombarding boron with alpha particles produces an unstable form of nitrogen with a half life of eleven minutes, while bombarding magnesium with alpha particles produces unstable isotopes of silicon and aluminium.

For all this work, Irène and Frédéric shared the Nobel Prize in Chemistry in 1935, the same year that Chadwick received the Nobel Prize in Physics. Their citation read 'in recognition of their synthesis of new radioactive elements'; his was 'for the discovery of the neutron'.

Cloud chamber photograph taken by Frederic and Irene Joliot-Curie, showing the conversion of a gamma ray (a very energetic form of electromagnetic radiation) into an electron-positron pair. Gamma rays do not leave tracks in a cloud chamber because they have no electric charge. But if they have enough energy, they can reveal themselves by converting into matter and antimatter, as in this case. Because they have opposite electric charge, the electron and positron (anti-electron) curl away in opposite directions in the cloud chamber's magnetic field.

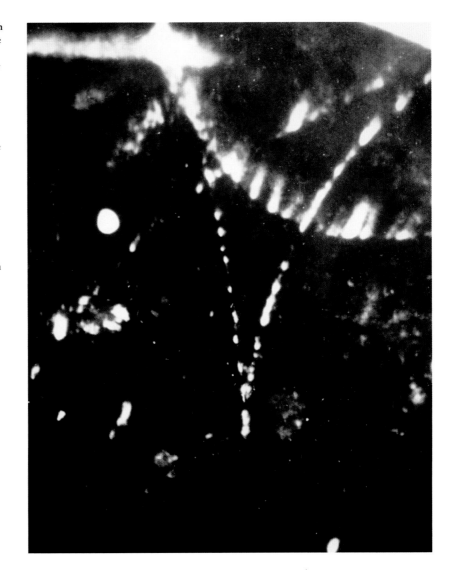

Looking ahead in his Nobel lecture, Frédéric Joliot-Curie accurately assessed some of the long-term implications of this work: '[If] we cast a glance at the progress achieved by science at an ever-increasing pace, we are entitled to think that scientists, building up or shattering elements at will, will be able to bring about transmutations of an explosive type, true chemical chain reactions. If such transmutations do succeed in spreading in matter, the enormous liberation of usable energy can be imagined. But, unfortunately, if the contagion spreads to all the elements of our planet, the consequences of unloosing such a cataclysm can only be viewed with apprehension.'[41]

Within ten years of those words being uttered, the first nuclear bomb was exploded.

THE CAT IN THE BOX

One of the most famous, and most misunderstood, 'thought experiments' involves a hypothetical cat that is neither dead nor alive, or maybe both dead and alive at the same time. It was put forward by the Austrian physicist Erwin Schrödinger in 1935, following discussions with Albert Einstein about what they both regarded as the absurdity of the accepted understanding of quantum physics.

In the 1930s, the accepted image of what goes on in the quantum world was known as the Copenhagen Interpretation, because its leading proponent was the Dane Niels Bohr. This interpretation sought to explain puzzles such as the dual wave-particle nature of things like electrons (see page 181), by saying that quantum entities become real only when they are observed. An experiment designed to find a wave will indeed find a wave, while an experiment designed to find a particle will indeed find a particle. But you can never see both aspects at once, and when it is not being observed a thing like an electron exists in a fuzzy, indeterminate state known as a superposition. The act of observing, or measuring, a quantum entity forces it to 'collapse' into one definite state. The rules of quantum behaviour worked out from experiments make it possible to calculate probabilities for the 'collapse' of the superposition into different realities, but you can never be certain what is going to happen.

Schrödinger sought to demonstrate the absurdity of this package of ideas by scaling it up from the level of atoms and electrons to the level of cats. No real cats were involved. The 'experiment' is purely a mental exercise. But it is still revealing.

Schrödinger asked us to imagine a cat shut up in a sealed room with an ample supply of air, food and water. (His original German paper referred to a 'chamber', but somehow in translation this has become known as a 'box', with unfortunate implications for the cat's comfort, and the whole thing is known as the 'cat in the box' experiment.) In its sealed chamber, the cat has all the essentials of a comfortable life. But there is also what Schrödinger called a 'diabolical device' which monitors the decay of some radioactive material. If the radioactive material spits out an alpha particle, the device will crack open a bottle of poison and the cat will die.

The rules of probability that apply in the world of quantum physics, and especially the Copenhagen Interpretation, make it possible, in principle, to set up such an apparatus in such a way that after a certain time there is an exact 50:50 chance that the decay has occurred and the alpha particle has been emitted. But there is an equal probability that nothing has happened. Nobody has looked at the experiment, so according to the Copenhagen

Erwin Schrödinger
(1887–1961).

Schrödinger's cat.

Interpretation the radioactive material is in a superposition, poised between decaying and not decaying. So, said Schrödinger, the detector must also be poised in a superposition, the bottle of poison is poised between being open and closed, and the cat is poised in a superposition of live and dead states at the same time. It is only when somebody opens the door of the room and looks in that the whole system collapses into one state, with either a live cat or a dead cat.

Schrödinger thought that it was absurd to suggest that the act of looking in the room could affect things in this way, and that therefore the Copenhagen Interpretation was wrong. Not only wrong, but ridiculous. This did not stop people from using the Copenhagen Interpretation (some people still do), because the equations work, whatever the interpretation you put on them. But the thought experiment generated a continuing debate among the theorists about where to draw the line between the quantum world and the everyday world. Can the detector be regarded as an 'observer' which causes the superposition to collapse? And how about the cat? Surely the cat knows if it is dead or alive, and can trigger the collapse without human help. Then there is the possibility that there is no collapse, because there is one world in which the cat lives and one where it dies. This is known as the Many Worlds Interpretation of quantum physics – 'many' because you need one world for each outcome of every quantum level experiment.

None of this affects the calculations. If you want to design, say, a quantum computer, the equations are the same whichever interpretation you favour. Most quantum physicists keep quiet about the interpretations and just do the calculations. But the puzzle posed by Schrödinger with his cat-in-the-box thought experiment is as puzzling today as it was in 1935, and hints that for all the success of the calculations we do not really know what is going on in the subatomic world.

Nº 76 FISSION GETS HEAVY

Although the experiments of John Cockcroft and Ernest Walton (see page 192) did involve 'splitting the atom', the atoms involved were very light. What most people think of today when atom splitting is mentioned is the kind of fission that occurs when the nucleus of a heavy element such as uranium splits into two or more fragments. In the 1930s, several teams made this happen in their laboratories without realizing what was going on. It was Otto Hahn, based in Berlin, who, together with his colleagues, not only triggered nuclear fission but explained what had happened in their experiment.

Hahn and his colleague Lise Meitner decided to follow up the discovery, made by Enrico Fermi and his team in Rome in 1934, that what seemed to be 'new' elements could be made by bombarding uranium with neutrons. Fermi thought that he had made elements with 93 and 94 protons in their nucleus, which he named ausonium and hesperium. Uranium has 92 protons in each nucleus, and comes in several different varieties (isotopes) including those with 143 and 146 neutrons per nucleus, dubbed uranium235 and uranium238. The German chemist Ida Noddack immediately suggested an alternative, that 'it is conceivable that the nucleus breaks up into several large fragments,' but nobody followed this up.

Hahn and Meitner, working with Fritz Strassman, set out to investigate Fermi's discoveries. Meitner had been working with Hahn since before the First World War, but before they could complete this new project, Meitner, an Austrian Jew, had to leave Berlin because of the risk of Nazi persecution. In the summer of 1938 she moved to Sweden. But she corresponded with Hahn about the progress of the experiments he continued to carry out with Strassman, and contributed a theoretical understanding of what was going on.

Before and after Meitner's departure, the team carried out experiments along the same lines as those of Fermi, which seemed to show the production of new 'transuranium' elements. But towards the end of 1938 Hahn discovered something unexpected. Chemical analysis showed that one of the 'new' elements being produced by bombarding uranium with neutrons was an isotope of barium – an atom with 60 per cent of the mass of a uranium atom. The analysis that revealed the trace of barium involved only a few thousand atoms, but Hahn was a superb chemist and when Meitner received a letter from him with news of the discovery, in December 1938, she had no doubt that he was right. Hahn himself was baffled. He wrote 'we are more and more coming to the awful conclusion that our Ra [radium] isotopes behave not like Ra, but like Ba [barium] … Perhaps you can suggest some fantastic explanation.' She could. Uranium nuclei had been split apart by the bombardment.

The explanation was that the nucleus of an atom is like a drop of liquid, held together by the strong nuclear force (see page 191), but with the positive charge

Lise Meitner and Otto Hahn, in their laboratory at Dahlem, Germany.

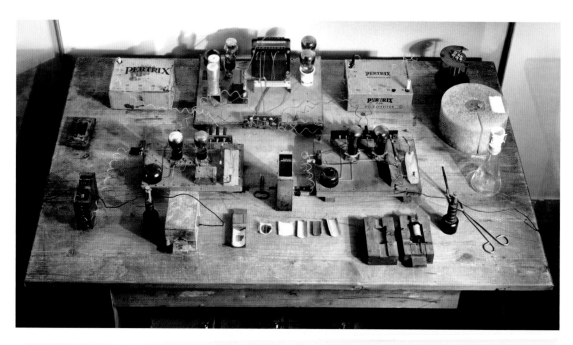

A museum display set up to illustrate Otto Hahn's workbench. It shows equipment used by Hahn in early nuclear fission experiments with Lise Meitner and Fritz Strassmann. The equipment includes batteries (Pertrix), amplifiers (three, one between batteries), automatic counters (lower left and centre), Geiger-Muller tubes (across bottom), and a paraffin block (circular, upper right) containing a neutron source (round) and uranium (rectangular).

of all the protons trying to blow it apart. In the case of uranium, with so many protons the electric repulsion is nearly enough to overcome the strong force, and if the drop is hit by a fast-moving neutron it can break into two or more droplets, which are then repelled from one another by the electric force. When uranium, with 92 protons, splits to produce barium, with 56 protons, the other 'droplet' is krypton, with 36 protons. Several neutrons are spat out in the process. The energy carried by these fragments turned out to be about 200 million electron Volts (200 MeV) in each fission.

Meitner was by now collaborating with her nephew Otto Frisch. Frisch was a nuclear physicist based in Denmark, but had been visiting her when Hahn's letter arrived. She calculated the masses of the two nuclei formed by the division of a uranium nucleus and found that their combined mass is less than the mass of a uranium nucleus by one-fifth the mass of a proton. In line with Einstein's famous equation $E = mc^2$, one-fifth of a proton mass is equivalent to 200 MeV. So everything fitted and uranium fission provided a potential source of energy for peaceful and military use.

As Hahn commented, the 'active breakdown products, previously considered to be transuraniums, were in fact not transuraniums but fragments produced by splitting.'[42] It was Frisch who suggested the name nuclear fission (in German, Kernspaltung) which appeared in a paper he published with Meitner in 1939. Hahn alone received the Nobel Prize for chemistry for this work. It was the prize for 1944, but held over until 1945. The citation read 'for his discovery of the fission of heavy nuclei'. This highlights the curious fact that one of the most important discoveries in physics was indeed made by chemical analysis of tiny traces of a substance.

Following the work of John Cockcroft and Ernest Walton (see page 192), the Hungarian physicist Leó Szilárd, then based in England, pointed out the possibility of a nuclear chain reaction. In such a process, if a neutron striking a nucleus caused it to release more neutrons, these neutrons in turn could trigger further nuclei to release neutrons in a self-sustaining and possibly runaway reaction. Little notice was taken of the idea until 1938, when Szilárd moved to New York. There, he heard about the work of Lise Meitner and Otto Hahn (see page 204) with uranium fission. He realized that it might be possible to produce a self-sustaining uranium fission reaction, which would release energy as the uranium nuclei 'split'. A small-scale experiment proved the idea in principle, and Szilárd organized a letter to the President, Franklin D. Roosevelt, warning of the possibility that the Germans might develop a nuclear bomb using this technique. (Albert Einstein was persuaded to sign the letter, although he had nothing to do with this work.)

When the United States entered the Second World War, the fear of being attacked with nuclear weapons lead to the creation of the Manhattan Project, to produce a nuclear bomb first. One of the steps towards the bomb involved the construction of a nuclear reactor, in which a pile (literally) of uranium could be made to produce energy, in the form of heat, without the reaction running away and making an explosion.

The person in charge of the construction of this experimental reactor was Enrico Fermi, who had moved to America because of the threat to his Jewish wife of the Fascist regime in Italy. It depended on having just the right amount of uranium in one place, with a way to control the number of neutrons taking part in the reaction. In a very small lump of uranium, most of the neutrons produced

Enrico Fermi (1901–1954).

by spontaneous fission escape from the surface and do not trigger the fission of other nuclei. In a very large lump of uranium, neutrons from fission that occurs inside the metal will be likely to hit other nuclei before they can escape, triggering more fission in a runaway process that causes an explosion. The reactor built under Fermi's direction, inside a disused squash court at the University of Chicago, struck a balance between these extremes.

The reactor was known as Chicago Pile-1, or CP-1. The pile was built on a square base but with a rounded top, and was made up from alternating layers of graphite blocks and graphite blocks containing lumps of uranium metal or uranium oxide. Graphite absorbs or slows down neutrons, so it helps to 'moderate' the reaction. Fermi and Szilárd had calculated that placing the uranium in blocks of moderating material, forming a cubical lattice of uranium, would give the best chance for a neutron from one uranium atom to encounter the nucleus of another uranium atom. The key control mechanism, however, was a rod made of cadmium, which absorbs neutrons, that could be moved in or out of the pile as required. With the rod inserted, neutrons would be absorbed and there would be no chain reaction; with the rod removed, the neutrons could fly freely to trigger fission and keep the reaction going.

Fermi calculated that a pile of these graphite blocks 56 layers high would contain enough uranium to produce a self-sustaining nuclear reaction. So CP1 was built 57 layers high. The finished reactor had cost roughly a million dollars to manufacture, and contained 771,000 pounds (350 tons) of graphite, 80,590 pounds (37 tons) of uranium oxide and 12,400 pounds (5.6 tons) of uranium metal. It was 25 feet wide and 20 feet high (8 metres wide by 15 metres high). At 3.36 pm local time on 2 December 1942, the cadmium rod was slowly withdrawn from the pile, allowing the neutrons produced by fissioning uranium nuclei to trigger the fission of other nuclei. The reactor operated as predicted for 28 minutes before the rod was pushed back in and the reaction shut down. It was the beginning of the so-called 'atomic' (actually nuclear) age.

The success of CP1 showed not only that a nuclear chain reaction could be produced artificially, but, even more importantly, that it could be controlled. The maximum power output from the reactor was only some 200 watts, enough to run an incandescent light bulb, but as Fermi later said, 'we all hoped that with the end of the war emphasis would be shifted decidedly from the weapon to the peaceful aspects of atomic energy' including 'the building of power plants'.[43] But he did not live to see the full fruition of this dream. Built under wartime emergency conditions, Chicago Pile 1 had no radiation shielding of any kind; on 28 November 1954, at the age of 53, Enrico Fermi died of stomach cancer, possibly caused by his work with radioactive material.

Artwork showing the moment when the world's first nuclear reactor, Chicago Pile-1 (CP-1), became self-sustaining.

THE FIRST PROGRAMMABLE COMPUTER

An experimental prototype of what became the world's first programmable computer (a primitive 'Turing machine') began operating at the end of 1943, and paved the way for several such machines to be fully operational by June 1944. The dates are important, because these machines played a key role in providing intelligence for the planners of the D-Day landings in Normandy in the Second World War.

The intelligence was gathered by code-breakers working at the British secret establishment at Bletchley Park. They had previously achieved great success in breaking the German Enigma code using less sophisticated machines, based on principles developed by Alan Turing. But by 1943 the Germans had improved their codes to the point where it was clear that a new kind of machine would be needed to tackle them.

Thomas Flowers, a Post Office research engineer who had already done some work for Bletchley Park, suggested that the solution to the problem would be to build a computer based on electronic valves (called 'tubes' in America), which control the flow of electricity through a system and glow like little light bulbs. Flowers had experience of working with valves because he had been involved in developing a valve-based telephone exchange in the 1930s. He had made the key realization that although these valves tended to break down very quickly when they were turned on and off, they had a much longer life if they were left on all the time, even when not in use.

The authorities at Bletchley Park were not convinced, and in February 1943 Flowers went back to his work at the Post Office's Dollis Hill research station, where the Director allowed him to work on the project semi-officially. But funds were so limited that Flowers paid for much of the equipment himself. The result was a machine using 1,600 valves, fed by a paper tape containing the message to be broken as a series of punched holes. Because of its size, it was dubbed 'Colossus'.

Colossus was tested at Dollis Hill in December 1943, then dismantled, taken to Bletchley Park in pieces in a series of lorries and re-assembled. There, it broke its first message on 5 February 1944. The code-breakers were astonished. Just a year after having the idea dismissed as pie in the sky, Flowers was told by the authorities to get at least one improved Colossus up and running by 1 June. He was not told why, but just met the deadline with a machine using 2,400 valves. On 5 June, Colossus II decoded a German message which revealed that the German high command had fallen for a deception plan suggesting that the imminent invasion would take place in the Pas de Calais, where Erwin Rommel was told to concentrate his forces. This intelligence was a key factor in persuading General Dwight D. Eisenhower, the Allied commander, to go ahead with the Normandy invasion on 6 June.

Colossus, the world's first electronic programmable computer, at Bletchley Park in Buckinghamshire.

Although the 'Colossi' (several more were built) were there primarily to crack codes, they had a much greater potential. Flowers had designed them so that they could be adapted ('programmed', we would now say) to carry out different roles by flipping various switches, and plugging the leads that connected the different parts of the machine in different configurations. This programming had to be done literally by hand, but it made the Colossi true computers in the modern sense of the word. In the mid1930s, Turing had shown mathematically, in a scientific paper entitled 'On Computable Numbers', that it is possible to build a machine that will tackle any problem that can be expressed in numerical (digital) terms. The same piece of machinery (what we now call hardware) could be made to do any possible task in accordance with the appropriate sets of instructions (what we now call software) expressed in binary code as a string of 1s and 0s. Such a machine, or computer, is now known as a universal Turing machine, or just a Turing machine, a term that embraces all modern computers, including all the ones in smart phones as well as the big machines used in tasks such as weather forecasting. The Bletchley Park researchers were well aware that the Colossi had the potential to manipulate all kinds of data, including storing pictures, word-processing, or anything that could be expressed in numbers. They had all the components of a genuine all-purpose calculating machine.

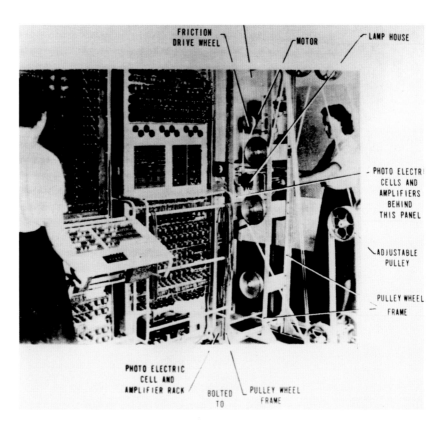

Colossus in operation.

All this was kept secret even after the war (on the direct instructions of Winston Churchill). As a result, an American machine dubbed ENIAC (from Electronic Numeral Integrator And Computer), which began operating in 1945, is still sometimes referred to as the world's first electronic computer. Like Colossus, it could be programmed using a plugboard system and, unlike Colossus, it was developed to provide weather forecasts and analyse wind-tunnel tests, among other tasks. But Colossus, and Flowers, were first.

N⁰· 79   DISCOVERING THE ROLE OF DNA

After Fred Griffith had discovered, in 1928, that one form of pneumococci could be 'transformed' into another by absorbing what seemed to be genetic material (see page 184), other researchers tried to find out just what it was that was being passed from one form of the bacteria to another. The key developments took place at the Rockefeller Institute in New York, in a laboratory headed by Oswald Avery, who had been working on pneumonia since 1913, and took a keen interest in Griffith's discovery.

In 1931, the Rockefeller team found that it wasn't even necessary to use mice in the transformation process. Just by growing R pneumococci in a Petrie dish also containing dead S pneumococci they were able to transform the live R type into the S type. This began a quest to identify the transforming agent. By using a process of alternate freezing and heating, the cells of a colony of Stype bacteria were broken apart so that their interior contents mixed with the other fragments of the cells in a liquid goo. By spinning test tubes containing this goo in a centrifuge, the solid pieces of cell-wall debris were forced to the bottom of the tubes, leaving above them a light liquid (the inner contents of the cells). Sure enough, the liquid from inside the cells was able to transform R-type cells into the S type.

All of this was established by 1935. At this point, Avery brought in a young researcher, Colin MacLeod, to work with him on an intensive investigation of the genetically active liquid from inside the cells. They were later joined by Maclyn McCarty. It took them nearly ten years to complete the project, not least because along the way they had to eliminate all the ingredients that were not causing the transformation, until they were left with the culprit. Paraphrasing the words Arthur Conan Doyle put into the mouth of his character Sherlock Holmes, once they had eliminated the impossible, what was left, however improbable it might have seemed at the start of the investigation, must be the answer.

It seemed at first that the most likely transforming agent might be protein, as proteins are very complex molecules that contain a lot of information. But when

Oswald Avery (1877–1955), who discovered that deoxyribonucleic acid (DNA) serves as genetic material.

the team treated the liquid derived from the cells of Stype bacteria with an enzyme that was known to chop protein molecules into little pieces (a protease), they found that this had no effect on the ability of the liquid to carry out the transforming process. Another possibility was that the effect was associated with the polysaccharides that coated the Stype bacteria. So Avery's team used an enzyme which broke the polysaccharides apart, but again with no effect on the transforming process. At this point, in a careful series of chemical steps, the team removed all traces of proteins and polysaccharides from their brew, and set about a painstaking chemical analysis of what was left behind. It had to be a nucleic acid (see page 170), revealed by the proportions of carbon, hydrogen, nitrogen, and phosphorus that it contained. Further tests revealed that it was DNA, not RNA.

The discovery was published in 1944, in the first of a series of scientific papers reporting the identification of the transforming agent as DNA. They stopped short of saying that DNA must be the material that genes are made of, although Avery did speculate about this possibility privately, including in a letter to his brother Roy, a bacteriologist. But the suggestion that DNA, not protein, carried the hereditary information stored inside cells was so shocking that the biological community at large did not immediately take it on board. They were still largely convinced that DNA was too simple a molecule to do the job, held back by the 'tetranucleotide hypothesis' (see page 172). In addition, as the active material inside the cells of bacteria is floating around loose inside the cell and not packaged into genes and chromosomes, many biologists at the time thought that it was too big a leap to jump from DNA as the transforming factor revealed by

Griffith's work, to DNA as the active component in true genetics. Nevertheless, the Avery–MacLeod–McCarty experiment attracted wide interest and stimulated more work by microbiologists and geneticists on the physical and chemical nature of genes. It is now recognized as the beginning of molecular genetics. If ever a team deserved the Nobel Prize, Avery, MacLeod, and McCarty did – but somehow, they were overlooked. It would be several years before the balance of evidence in favour of DNA as the genetic material became overwhelming, thanks to another brilliant experiment (see page 219).

Nº. 80 JUMPING GENES

Until the late 1940s, 'everybody knew' that genes were stable entities strung out along a chromosome like beads on a wire. That image was challenged by Barbara McClintock, working at the Cold Spring Harbor Laboratory in New York the United States; but it took a long time for the significance of her experiments to be widely recognized.

McClintock started her research at Cornell University, New York, in the 1920s, using studies of maize plants (sweetcorn) to build upon the pioneering work of Thomas Hunt Morgan with fruit flies (see page 173), in an echo of the way in which Gregor Mendel had studied heredity using pea plants (see page 134). She developed techniques to stain the cells of maize plants so that the ten chromosomes they contained could be clearly distinguished using a microscope, and, with the assistance of Harriet Creighton, she was able to show that specific changes in the chromosomes were associated with specific changes in the plant itself – in the phenotype. It was possible to identify individual genes within the

Barbara McClintock at work.

The kind of maize studied by geneticists such as Barbara McClintock has multi-coloured cobs.

chromosomes, and to compare the chromosomes of one plant, or one cell, with those of another. In one example, a maize strain being studied carried chromosomes that were associated with either light or dark kernels in the corn cobs. When the cells were stained and studied under the microscope, the researchers noticed that the difference was associated with a difference in chromosome number 9, with the chromosome in the dark variety having a visible knob which was not present in the equivalent chromosome of the other variety.

These results were published in 1931, but the experiments took a long time to develop further, for the same reason that Mendel's experiments took a long time – the slow rate at which plants reproduce, compared with things like the *Drosophila* fruit fly. The advantage is that instead of having to catch flies an eighth of an inch long and look at the colour of their eyes, you can simply peel back the leaves of the corn cobs and see the pattern of coloured seeds lined up for inspection. (Wild maize, unlike the kind sold in supermarkets, has multi-coloured cobs.) McClintock patiently continued her studies, moving to Cold Spring Harbor in 1941, and made her most important discoveries there in the years following 1944.

In order to study the genetics of maize, McClintock did not need to know whether it was protein or DNA that carried the genetic information. All she had to know was that the information was carried on the chromosomes, and that each chromosome is made up of separate genes that carry specific instructions for the life processes of the plant. McClintock was studying how mutations (changes in the genes) affect the pattern of pigmentation in the corn – how the genotype affects the phenotype. In some plants, she saw splashes of the 'wrong' colour on some of the leaves, such as a streak of dark green on a light green leaf. Because the leaf grows by the repeated division of cells, she could trace this mutation back to the very cell, at the start of the streak of colour, in which the mutation had occurred.

In some cases, these multicoloured patches developed at a different rate from the rest of the plant, either growing more quickly or growing more slowly than their neighbours. After years of painstaking research, involving generations of controlled crossbreeding of her plants, McClintock was convinced that what she was seeing was evidence that the behaviour of some genes is controlled by other genes. In this specific example, one of the control genes sat next to the gene responsible for colouring the leaf, on the same chromosome, and turned it on or off. But a second control gene affected the speed with which this process occurred, and that gene did not have to be on the same chromosome. Finally, having identified these genes she found that the second gene does not even have to stay in the same place. It could get shunted about during the process of cell division, so that by comparing different cells it looked as if it had moved from one place to another within a chromosome, or had even 'jumped' from one chromosome to another, a process she called transposition. This flew in the face of the established idea of the genome as a fixed set of instructions passed unchanged from one generation to the next.

McClintock's summarized her work in a scientific paper titled, 'The origin and behavior of mutable loci in maize,' published in 1950. It eventually changed the way scientists think about genetics and inheritance and opened the way to an understanding of how genes regulate the production of proteins in the body, and to developments in genetic engineering. In 1983, at the age of 81, McClintock belatedly received the Nobel Prize in Physiology or Medicine 'for her discovery of mobile genetic elements'.

Nº. 81   THE ALPHA HELIX

One of the most important 'experiments' in the history of molecular biology consisted simply of folding up a long strip of paper to make kind of concertinaed snake. The folding was carried out by the American chemist Linus Pauling, in 1948, who did it to explain the X-ray diffraction patterns produced by certain kinds of protein, which had been puzzling him (and others) for years.

Proteins come in two main varieties, both based on long chains, known as polypeptides. In one variety, the fibrous proteins, the molecules largely retain the long, thin structure you associate with a chain; in the other, the globular proteins, the chain is folded in upon itself, screwed up to make a ball. Fibrous proteins are an important structural material – the basis of things such as hair, feathers, muscles, silk, and horn. Globular proteins are workers, like the haemoglobin molecules that carry oxygen around in your blood.

The first X-ray diffraction images of fibrous protein were obtained by William Astbury, at the University of Leeds, in the 1930s. He was working with

OPPOSITE **Computer-generated image of a strand of the protein collagen.**

keratin, which is found in wool, hair and fingernails. The images showed a regular, repeating pattern, which suggested that the protein had a simple structure, but there was not enough information to work out exactly what that structure was. Pauling, who was the first person to work out the rules of quantum chemistry, and would write a definitive book on the subject,[44] was intrigued, and spent the summer of 1937 trying (unsuccessfully) to find a way of coiling a polypeptide chain to match the data. He decided that it would be necessary to go back to basics, looking at the amino acids that are the links in a polypeptide chain (see page 195) and trying to work out how they fitted together. But with other work and the Second World War intervening, it was a long time before he got to grips with the problem.

The first step was to study X-ray diffraction photographs of individual amino acids, which Pauling did at Caltech (the California Institute of Technology) with Robert Corey. The key thing that they discovered is that although many chemical bonds allow the atoms or chemical units on either side of the bond to rotate, the peptide bond between carbon and nitrogen (which gives polypeptides their name) is locked by a phenomenon known as quantum resonance, so that a chain containing these bonds cannot rotate around them. This part of the chain is held rigid. But still Pauling couldn't work out how to fold the chain up to match Astbury's photographs.

Although based at Caltech, Pauling spent some time in 1948 as a Visiting Professor at the University of Oxford, in England. In the spring of that year, he caught a bad cold, and spent a couple of days in bed reading science fiction and detective stories. It was when he got bored with these that he carried out the 'experiment' which revealed the structure of keratin.

Pauling took a strip of paper, and drew out along it a representation of a long polypeptide chain. He knew enough to get the distances between the various components roughly correct from memory, and the angles that the different units should make with each other. But there was no way he could make the chain fit along the straight, flat piece of paper. One key angle (the same link repeated at different places along the chain) always came out wrong, and it could not be altered because of the rigidity imposed by the CN quantum resonance. So he tried creasing the paper and folding it along repeated parallel lines to make the correct angle, 110 degrees, at each of these links in the chain. The creased strip of

Linus Pauling (1901–1994), with a rope showing how helices coil around one another in some proteins.

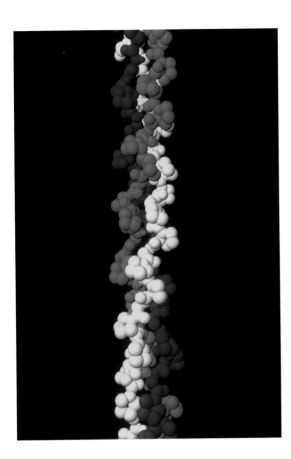

paper now had a roughly helical shape, a corkscrew of repeating linkages spiralling through space, in which some of the atoms were now located in just the right way for a quantum phenomenon known as hydrogen bonding to help to hold it together.

Back in the USA, further X-ray studies confirmed the correctness of this idea, and Pauling's team published a series of seven scientific papers in 1951 describing the structure of many fibrous proteins in terms of what he dubbed the alpha helix. But the discovery itself was almost less important than the way the discovery had been made. Pauling's breakthrough set people thinking about the role of helices in biological molecules, and opened their eyes to the value of starting from the bottom up, with the basic building blocks of biological material, to work out how they fitted together by building models – even if, as in this case, the model was only a folded strip of paper. Within a couple of years, this approach would harvest the biggest prize in molecular biology, the structure of DNA (see page 224).

Nᵒ. 82   A BLEND OF DNA

Even in 1951, is spite of the work of Oswald Avery and his colleagues (see page 213), it was still widely thought that genetic information was carried by proteins, not by DNA. But then came the experiment that persuaded even the doubters that DNA is 'the' life molecule.

The experimental steps along the road that led to an understanding of DNA used a succession of smaller and more rapidly reproducing organisms. Gregor Mendel worked with peas; Thomas Hunt Morgan worked with fruit flies; and Avery's team worked with bacteria. The final step was taken using the smallest entities that carry genetic material: viruses.

Viruses are little more than bags of protein, far smaller than a bacterium, filled with genetic material. When a virus attacks a cell, it injects this genetic material into the cell, where it hijacks the machinery of the cell and uses it to make copies of the virus out of the chemical material inside the cell. Then, the cell bursts open, releasing the copies of the virus to repeat the process. At the beginning of the 1950s, Alfred Hershey and Martha Chase, working at the Cold Spring Harbor Laboratory, developed a neat experiment which showed definitively that it was DNA that carried these instructions into the cell being attacked.

They worked with viruses known as bacteriophage (sometimes shortened to phage; from the Greek -*phagos*, to devour), because they 'eat' bacteria. The idea behind the experiment was based on growing phage in a medium that contained radioactive isotopes of either phosphorus (phosphorus-32) or sulphur (sulphur-35), then using the radioactive phage to attack a colony of non-radioactive bacteria. The sulphur and phosphorus isotopes have different and distinctive radioactive signatures, and they can be traced through the cycle of

A bacteriophage virus infects a bacterium and harnesses its DNA (deoxyribonucleic acid) to make numerous replications of itself. These replicants go on to infect other bacteria.

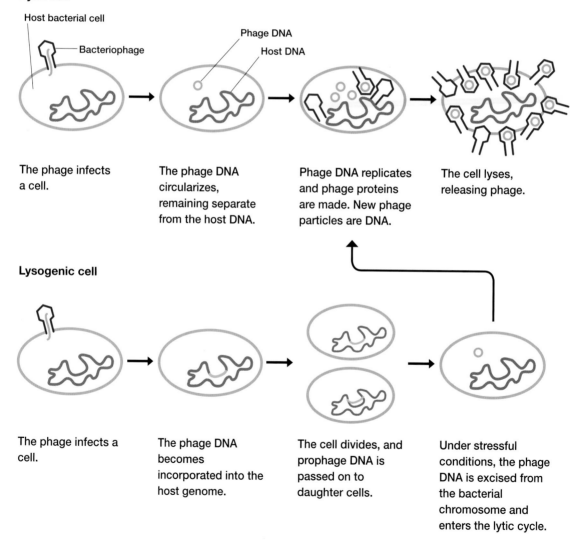

Lytic cell

Host bacterial cell

Bacteriophage

Phage DNA

Host DNA

The phage infects a cell.

The phage DNA circularizes, remaining separate from the host DNA.

Phage DNA replicates and phage proteins are made. New phage particles are DNA.

The cell lyses, releasing phage.

Lysogenic cell

The phage infects a cell.

The phage DNA becomes incorporated into the host genome.

The cell divides, and prophage DNA is passed on to daughter cells.

Under stressful conditions, the phage DNA is excised from the bacterial chromosome and enters the lytic cycle.

infection and reproduction from one generation of phage to the next. Bacteria were grown in colonies laced with one or other of the radioactive isotopes before phage was added. The phages in the next generation took up the radioactive material and were then used to infect non-radioactive bacteria. The point of all this is that phosphorus is present in DNA, but not in protein, while sulphur is present in protein but not in DNA. Everywhere the team detected phosphorus they would be tracing the path of DNA, while everywhere they detected sulphur they would be tracing the path of protein.

Martha Chase (1928–2003).

The snag was, after the radioactive phage had done their work in the culture of bacteria, what was left was a mass of cells filled to bursting point with new viruses but with discarded phage husks still attached to the hijacked bacterial cells. The culture still contained both kinds of radioactive isotope. Somehow, Hershey and Chase had to separate out the leftover debris from the original generation of phage from the new viruses manufactured inside the bacteria. The difficulty was solved when a colleague lent them an ordinary kitchen utensil known as a Waring Blender.

On a low setting, the blender provided just enough agitation to shake the empty phage bags loose from the cells they had infected, without smoothing everything out into an amorphous goo. After this agitation, the mixture was whirled in a centrifuge, where the bacterial cells, full of new virus, fell to the bottom and could be extracted, while the husks of the old phage were left behind. When the two components were analysed, the results were persuasive. Radioactively labelled DNA was found in the cells (in other words, in the new generation of viruses), while radioactively labelled protein was found in the leftover husks. It was DNA, not protein, that was passed from one generation to the next, although in the scientific paper announcing their results the team cautiously concluded only that 'this protein probably has no function in the growth of intracellular phage. The DNA has some function'.

The success of this superficially simple experiment owed a great deal to the expertise of Martha Chase, although officially she was 'only' the assistant to Alfred Hershey. Waclaw Szybalski, another Cold Spring Harbor biologist, later recalled: 'Experimentally, she contributed very much. The laboratory of Alfred Hershey was very unusual. At that time there were just the two of them, and when you entered the laboratory there was absolute silence and just Al directing the experiments by pointing with his finger to Martha, always with a minimum of words. She was perfectly fitted to work with Hershey.'[45]

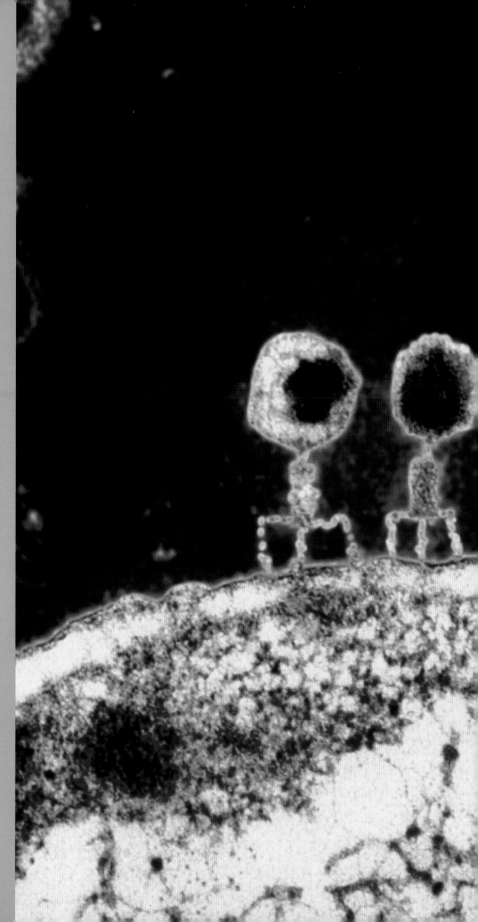

Coloured Transmission Electron Micrograph (TEM) of T-bacteriophage viruses attacking a bacterial cell of Escherichia coli. Seven virus particles are seen (blue), each with a head and a tail. Four of these are 'sitting' on the brown bacterial cell, and small blue 'tails' of genetic material (DNA) are seen being injected into the bacterium.

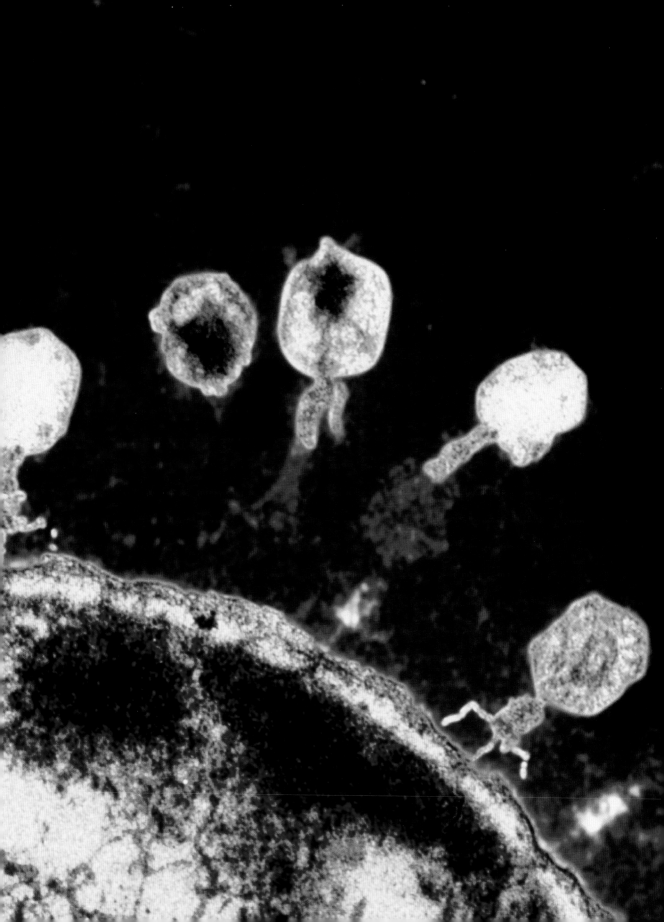

It was now clear that the protein provided the structural material in the bacteriophage, while the DNA carried the genetic information. The results were published in 1952, and the work became known as 'the Waring Blender experiment'. After this, hardly any biologists believed that the genetic material could be anything but DNA, and the stage was set for the investigations that would reveal the structure of DNA itself.

Nº 83 THE DOUBLE HELIX

The structure of DNA, the molecule of life, was revealed by experiments carried out by a team working at the Medical Research Council's Biophysics Research Unit at King's College, London. The head of the unit, John Randall, was one of the first people to accept the evidence that genetic information is carried by DNA, and in 1950 he assigned a research student, Raymond Gosling, to work with Maurice Wilkins to obtain X-ray diffraction images of DNA.

Working in a basement at King's, Gosling was able to crystallize DNA – the first person to crystallize genes. He knew that it was then just a matter of time before the structure of the molecules would be revealed. Under Wilkins' supervision, Gosling obtained the first X-ray diffraction images of crystalline DNA. These were presented by Wilkins at a meeting in Naples, where the American James Watson was in the audience and realized their significance.

Back at King's, the team were joined by Rosalind Franklin, who already had considerable expertise at X-ray diffraction work. Randall was particularly eager to bring her on board, because he doubted that Wilkins and Gosling had the expertise to solve the problem of working out the structure of DNA from the spotty diffraction images. Armed also with new equipment, Franklin and Gosling obtained better images, and discovered that there are two forms of crystalline DNA. When it is wet, it forms a long, thin fibre, but when it is dry, it is short and fat. These became known as the 'B type' and 'A type', respectively. Because of the humidity inside cells, the B type was expected to be more like the DNA found in living things.

The X-ray diffraction photographs hinted at the structure of the molecules of DNA, but a great deal of analysis was required to work out the positions of the atoms in the molecules. Franklin, with Gosling, concentrated on solving this puzzle for the A type, while Wilkins focused on the B type. There was some crossover between their work, but not as much as there should have been, because of a personality clash between Franklin and Wilkins. By the beginning of 1953, it was clear that both forms of DNA were based on a helical structure, and Franklin prepared two scientific papers suggesting a double-helical structure of the A type DNA, which were submitted to the journal *Acta Crystallographica* in March that year. Just at that time, she moved from King's to Birkbeck College, also in London. Also just at that time, the structure of the B type DNA had been determined by another team, in Cambridge.

In January 1953, on a visit to King's, Watson, who was now based in Cambridge, had been shown a print of the best X-ray diffraction image of the B type DNA. It had actually been taken in May 1952, by Franklin and Gosling, but was passed on without their knowledge. The image, known as Photo 51, showed the X-ray diffraction pattern of DNA in the highest quality available at the time and clearly shows a cross-shaped pattern that could only result from a helical structure. If one experiment could be said to have unlocked the secret of DNA, it was the one that produced this image.

Back in Cambridge, Watson and his colleague, Francis Crick, who had together been puzzling about the structure of DNA for some time, tried building models of the molecule that would fit the image, using Pauling's bottom-up approach (see page 217). This quickly led them to the discovery that everything fitted together if a DNA molecule consisted of two strands twined around each other in a double helix, with the bases on the inside, so that the bases on one strand linked with the bases on the other strand like the steps on a spiral staircase. Adenine always links with thymine, and cytosine always links with guanine. So the strands are like mirror images of each other, and if they unravel each lone strand can build a new double helix by adding the appropriate units to build up the other strand.

LEFT **Rosalind Franklin (1920–1958).**

RIGHT **'Photo 51.' X-ray diffraction photograph of DNA (deoxyribonucleic acid), obtained in May 1952 by King's College London researchers Rosalind Franklin and Raymond Gosling (1926–2015).**

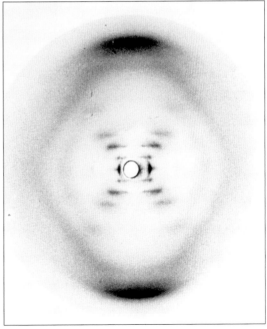

James Watson (b.1928), left, and Francis Crick (1916–2004), with their model of part of a DNA molecule in 1953.

The structure also carries much more information than a boring repetition of a four-letter word (see page 265). The A, T, C and G can occur along the strand in any order, such as AATCAGTCAGGCATT …, like a message in a four-letter code. This can carry a great deal of information, just as the two-letter Morse code or binary computer code can convey a great deal of information. Crick and Watson completed their model building on 7 March 1953, and sent a paper off to *Nature*.

In 1962, Crick, Watson and Wilkins shared the Nobel Prize in Physiology or Medicine for this work. Franklin, who had died of cancer in 1958, could not share the honour as Nobels are never given posthumously.

MAKING THE MOLECULES OF LIFE

A s it became clear that the molecules of life obey the same chemical rules as other molecules, the question of how living material had evolved from non-living material moved from the realm of philosophy into the realm of science. As early as 1922, the Russian chemist Alexander Oparin jumped off from the then recent discovery of methane in the atmospheres of the giant planets to suggest that the early Earth had possessed a strongly 'reducing' atmosphere containing methane, ammonia, water, and hydrogen, which had provided the raw materials of life molecules, such as proteins, thanks to the input of energy from lightning and ultraviolet radiation from the Sun. Similar ideas were developed independently in the 1920s by the British polymath J. B. S. Haldane, who coined the term 'primordial soup' to describe the conditions in the ocean of the young Earth.

Miller-Urey experiment.

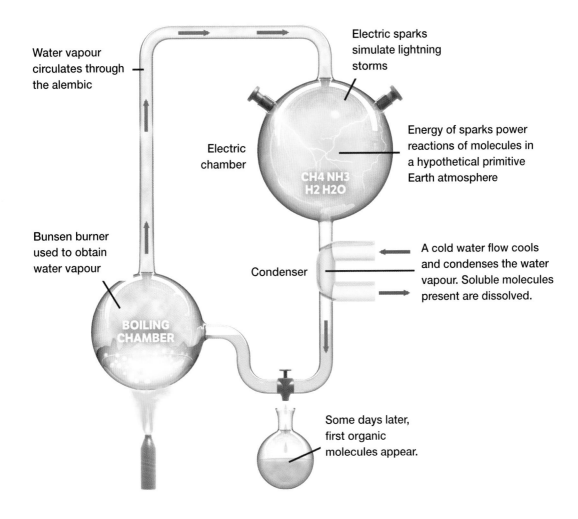

Water vapour circulates through the alembic

Electric sparks simulate lightning storms

Electric chamber

Energy of sparks power reactions of molecules in a hypothetical primitive Earth atmosphere

CH4 NH3 H2 H2O

Bunsen burner used to obtain water vapour

BOILING CHAMBER

Condenser

A cold water flow cools and condenses the water vapour. Soluble molecules present are dissolved.

Some days later, first organic molecules appear.

Both of them, in fact, were thinking along the same lines as Charles Darwin, who had written in a letter to Joseph Hooker in 1871: 'If (and oh! what a big if!) we could conceive [of] some warm little pond, with all sorts of ammonia and phosphoric salts, light, heat, electricity *etc.* present, that a protein compound was chemically formed ready to undergo still more complex changes.'[46]

In the 1950s, the American chemist Harold Urey (who had received the Nobel Prize in 1934 for his discovery of deuterium) was working at the University of Chicago. He gave a lecture on the Oparin-Haldane hypothesis, which encouraged a graduate student, Stanley Miller, to ask if he could work on the task of constructing a suitable experiment to test the idea, under Urey's supervision. The resulting Miller-Urey experiment was a simple closed system in which a 5-litre glass flask containing methane, ammonia, water vapour, and hydrogen was energized by electric sparks to simulate lightning. A half-litre flask half full of steaming hot water provided water vapour to the mixture, with the hot gases leaving the first flask being cooled and circulated back over the boiling water into the flask. Any liquid produced could be caught in a trap, rather like the U-bend under a sink, and collected. The experiment could be run continuously for a week or more before the products were collected and analysed.

Within a day, the liquid held in the trap had turned pink. After a week of operating the experiment, the pink liquid was drained off and Miller found that between 10 and 15 per cent of the carbon present in the original mixture of gases had been incorporated into organic compounds, which included 13 of the 22 amino acids which themselves form the building blocks of proteins in living organisms. He had not created life, but he had produced what were regarded as precursors of life.

The results of the first version of the experiment were published in the journal *Science* in May 1953, and Miller got his PhD. He continued with research building on this work right up until his death in 2007, using improved equipment and more sophisticated chemical analysis technique, to produce a wide variety of complex organic molecules. By that time, however, it was thought that the atmosphere of the early Earth had probably not been the mixture of gases he had originally assumed, but was dominated by carbon dioxide, with gases such as nitrogen and sulphur dioxide released from volcanoes. But similar experiments with this kind of mixture of gasses, plus water vapour, give similar results to the Miller-Urey experiment.

The key discovery of the original experiment is that relatively simple raw materials plus a supply of energy can produce life molecules. It may not even be necessary for this process to occur on Earth. Astronomers have now detected, using spectroscopy, many complex organic molecules in clouds of gas and dust in space, including things such as formaldehyde and propylene, and a molecule known as iso-propyl cyanide, which has a structure similar to amino acids. And a

Harold Urey (1893–1981) in his laboratory.

meteorite that fell near Murchison, Victoria, in Australia on 28 September 1969, was later found to contain many amino acids, including 19 that are found in living things. It now seems likely that complex molecules of this kind, perhaps even including amino acids themselves, were brought down to Earth by comets soon after the planet formed, lacing Darwin's 'warm little pond' with compounds at least as complex as those produced by the Miller-Urey experiment. As Miller himself once said, 'The fact that the experiment is so simple that a high school student can almost reproduce it is not a negative at all. That fact that it works and is so simple is what is so great about it.' If a high-school student can manufacture the precursors of life in a week, it is no surprise that the Universe has been able to manufacture life in a few billion years.

MASERS AND LASERS

M asers and lasers are among the most ubiquitous inventions in modern society. They are used in telecommunications, supermarket bar-code readers, Bluray and CD players, fibre-optic communications and in many other applications. The idea behind them goes back to an insight from Albert Einstein in 1917, but the key experiments proving that the idea could be turned into practice were carried out by Charles Townes and his colleagues at Columbia University in 1953.

Einstein's insight is another example of the curious wave-particle duality of light and other electromagnetic radiation. He realized that if an atom had an electron in a so-called 'excited' state, with extra energy, then a photon with a certain wavelength passing by could make the electron fall down into a lower energy state, releasing another photon with the same wavelength as the first one. There would be two photons instead of one. This became known as stimulated emission of radiation, but for a long time it was regarded as a quantum-mechanical curiosity with no practical value.

In 1951, however, Townes realized that with the improvements in technology developed for radar work in the Second World War it ought to be possible to apply Einstein's idea to create intense beams of radiation at microwave wavelengths (short-wave radio wavelengths). If a whole array of atoms could be prepared in the excited state, then in a process like that of a runaway nuclear chain reaction (see page 207), one photon could trigger emission from one atom, providing two photons to trigger two more atoms, then four, eight, sixteen, and so on. Crucially, all these photons would have

LEFT **Nikolai Basov (1922-2001) with a maser, photographed at the Lebedev Physics Institute, USSR Academy of Sciences, Moscow, Russia.**

RIGHT **Alexander Mikhailovich Prokhorov (1916–2002).**

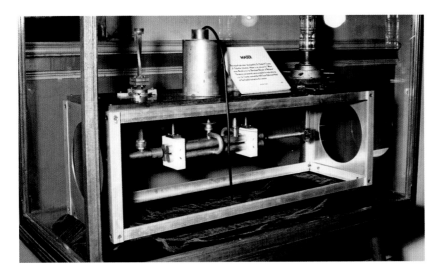

Townes-Gordon-Zeiger maser, on display at the Franklin Institute, Philadelphia, USA. Invented by Charles Townes (1915–2015), his postdoctoral student Herbert J. Zeiger (1925–2011) and his graduate student James Gordon (1928–2013).

exactly the same wavelength, marching in step and never cancelling one another out. This is known as coherent radiation. Townes called the process microwave amplification by stimulated emission of radiation, which got shortened to the acronym MASER, and then became a new word: maser.

The same theoretical idea was also developed by Nikolay Basov and Aleksander Prokhorov at the Lebedev Institute of Physics at about the same time. But neither team knew about the other group's work until after Townes, working with James Gordon and H. J. Zeiger, had constructed the first working maser.

In the key experiment, ammonia gas was radiated using microwaves with a frequency of just under 24,000 million cycles per second (24 GigaHertz). This put some of the molecules into the required excited state. The gas then passed through a device that used electric fields to extract most of the remaining unexcited molecules, which have slightly different electrical properties to the excited molecules. What was left was a gas in which most of the molecules were in the excited state – this is known as a population inversion, because it is the opposite of the usual state of affairs, and the whole process of creating a population inversion is known as 'pumping' the molecules. It was then possible to trigger the maser process, which produced a coherent amplified beam of pure radiation at the same frequency (24 GHz) as the incoherent radiation originally used to excite the molecules. This is equivalent to a wavelength of just over 1 centimetre.

The amplification took place in a cavity with carefully chosen dimensions (in effect a little metal box), so that photons could not escape but were bounced back to trigger more emission. Excited molecules passed through the cavity in a continuous stream, to keep the process going. The experiments of 1953 paved the way for the construction of the first fully functioning ammonia maser in 1954; its power output was just 10 nanowatts (ten billionths of a watt), but it proved that the trick worked.

With the principle established, other researchers were quick to develop a variety of masers and to scale up the energies involved. Although they knew that the same physical processes ought to work for light as well, it was much harder to work with light because it has much shorter wavelengths than the microwave radiation used in the ammonia experiment. But by 1960 the first system using light had been constructed. It was originally called an 'optical maser,' but such devices quickly became known as lasers (with the word 'light' replacing the word 'microwave' from MASER). In 1964, Townes, Basov, and Prokhorov shared the Nobel Prize in Physics 'for fundamental work in the field of quantum electronics, which has led to the construction of oscillators and amplifiers based on the maser-laser principle'.

A year later, in 1965, a strong source of microwave radiation from space was identified as the result of a naturally occurring maser involving the molecular radical OH in an interstellar gas cloud. Since then, at least a dozen other astronomical maser molecules have been identified, including water, ammonia, hydrogen cyanide, and methanol. It is not yet clear how these molecules are being 'pumped' into their excited states.

NO. 86 MAGNETIC STRIPES AND SEA-FLOOR SPREADING

The idea that continents move about on the surface of the Earth (that is, continental drift, or plate tectonics) is now one of the most firmly established scientific facts. But the theory only became respectable in the 1960s, when a variety of evidence in support of what had previously been speculation became available. One of the key pieces of evidence came from studies of the magnetism of the ocean floor; the experiment that set the scene was carried out in the second half of the 1950s.

At that time, at the height of the Cold War, the United States military were concerned about the possibility of Soviet submarines hiding on the sea bed near the west coast of North America. So in 1955 the US Navy funded a survey of part of the sea floor in that part of the world. The scientists carrying out the survey were told that they could also do other experiments provided this did not interfere with the survey, and they decided to study the magnetism of the sea floor. This involved towing a sensitive instrument known as a magnetometer behind the ship in a torpedo-like container.

The instrument measured how the overall magnetic field of the Earth changes from place to place. Where the underlying rocks were magnetized in the same way as the overall field, the measured field was a little stronger than average, and where the rock magnetism was aligned the opposite way to the Earth's overall field, the measured field was a little weaker. The researchers weren't looking for anything in particular, just taking advantage of the opportunity to look for anything that might be interesting.

To their surprise, they found something very interesting indeed. When the data were analysed, they showed a series of magnetic stripes, more or less parallel to one another. In one stripe, the rock magnetism is lined up in one direction, in the next stripe in the opposite direction, then once again in the first direction, and so on. The results were published in 1958, stimulating more detailed studies of larger regions of the sea floor. These showed that there is a symmetry in the stripes on either side of great ridges, underwater mountain ranges, that run down the middle of the oceans. The pattern on one side of the ridge is a mirror image of the pattern on the other side of the ridge.

The explanation came from studies of rock magnetism on land, which show that the entire magnetic field of the Earth reverses from time to time, with north and south magnetic poles being swapped. This does not involve the solid Earth toppling over in space, it is purely a magnetic effect. And when molten rock welling up from inside the Earth escapes from volcanoes and sets, it freezes in the direction of the magnetic field at the time.

This established that the sea floor itself, which is much thinner than the crust that makes up the continents, is growing, spreading outward from the volcanically active ocean ridges where lava escapes from cracks in the Earth's crust. The lava laid down either side of the ridge picks up the Earth's magnetism at the time and freezes it in. Then, when the field reverses the next strip of new ocean floor picks up the opposite magnetism. In the mid1960s, the US research ship *Elanin* surveyed an east-west path 4,000 kilometres in length, to the south of Easter Island, with the mid-Pacific ridge at its centre. When the magnetic stripes from the survey were plotted on a chart, the paper could be folded along the line representing the ridge, so that the two patterns sat exactly on top of each other, demonstrating the symmetry of the process.

But this does not mean that the Earth is getting bigger. The new crust being manufactured at ocean ridges spreads out on either side, pushing the continents apart. Continents do not move through the oceanic crust like an icebreaker plowing through ice but are carried on the backs of the plates (which give plate tectonics its name), pushed about like the ice flows being jostled on the surface of the water by currents underneath. At the edges of oceans there are places where the thin oceanic crust is pushed back down under the thicker continents and destroyed. This is responsible for the activity of volcanic regions such as the Japanese islands. And where thicker continental crust and thin oceanic crust are pushed together by continental drift, great mountain ranges, such as the Andes, can be built up as the crust crumples.

There are other, more subtle, processes involved in the way continents move about on the surface of the Earth, but sea-floor spreading, revealed by magnetic stripes on the ocean floor, is still the key feature.

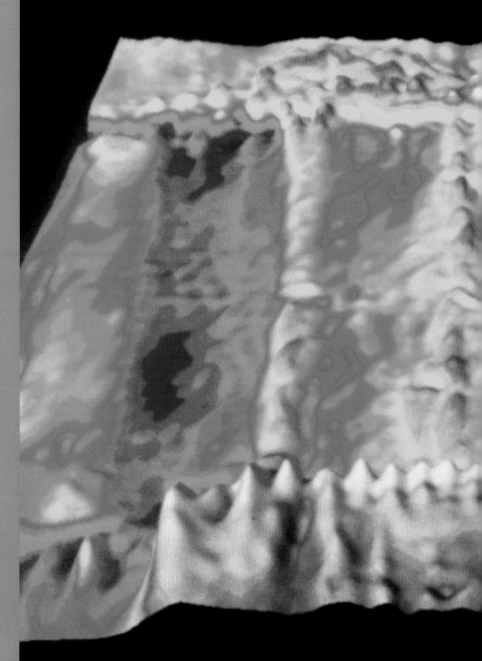

Image of the ocean floor showing part of the East Pacific Rise, an ocean spreading ridge. The vertical stripes of colour show changes in the Earth's magnetic field over the past few million years.

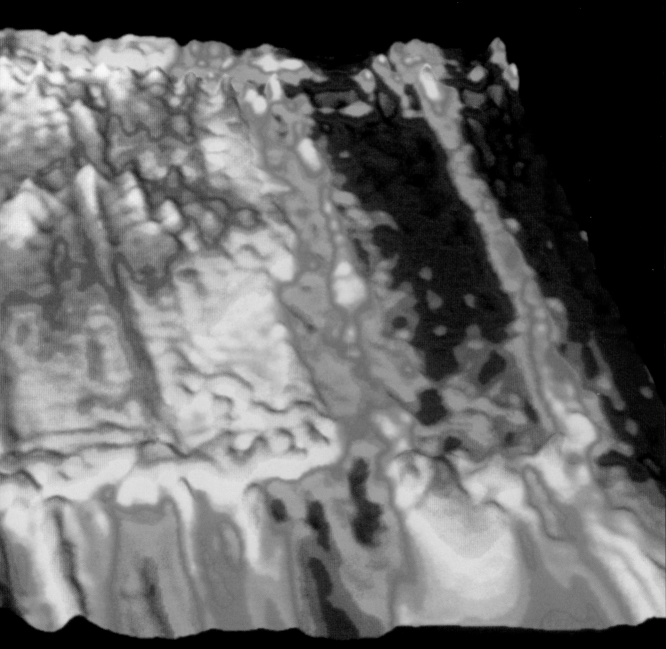

DETECTING THE GHOST PARTICLE

I n 1930, the Austrian physicist Wolfgang Pauli suggested that energy which seemed to be being lost during certain particle interactions was actually being carried away by a ghostly particle, which became known as the neutrino. The hypothetical particle had to have a very tiny mass (much less than that of an electron) and no electric charge, making it very difficult to detect. It would travel not just through space but through solid matter at almost the speed of light. But if it did not exist, the law of conservation of energy, which is one of the most fundamental laws of science, would have to be abandoned.

Theory said that apart from gravity, neutrinos only interact with matter through the so-called weak nuclear force, which is very weak indeed. If a beam of neutrinos like the ones thought to be produced by the nuclear reactions going on inside the Sun were to travel through a wall of solid lead for 3,500 light years, only half of them would be stopped along the way. Pauli himself thought that he had done 'a frightful thing', by proposing a particle that could never be discovered. He considered it so unlikely that any experiment would ever detect neutrinos directly that he offered a case of champagne to any experimenter who successfully took up the challenge.

But the development of nuclear reactors stimulated by the Second World War (see page 207) offered experimental possibilities undreamed of in the early 1930s. To trap one neutrino is almost impossible. But if you have a lot of neutrinos and a large enough detector you might hope to see the effects of a few of them interacting with the atoms in your detector.

The challenge was taken up by Clyde Cowan and Frederick Reines in the 1950s. Their detector was simply a tank of water (holding about 1,000 pounds, or 400 litres, of liquid) placed next to the Savannah River nuclear reactor in the United States. It was calculated that 50 trillion (5×10^{13}) neutrinos should be passing through every square centimetre of the side of the tank every second, and that, in the course of an hour, one or two of these neutrinos should be captured by a proton (a nucleus of hydrogen) in the water. This would convert the proton into a neutron and release a positron, the positively charged antiparticle counterpart to the electron. It was the positrons that Cowan and Reines set out to detect in their experiment. Each positron very quickly meets an electron, when the pair annihilate, emitting a pair of gamma rays with very distinctive properties.

Hints of the anticipated 'neutrino signal' came in 1953, and full confirmation that Pauli's idea was correct came in 1956. Cowan and Reines sent Pauli a telegram telling him the news, and

OPPOSITE **The giant tank of dry-cleaning fluid used in the Homestake mine solar neutrino detector.**

BELOW **Raymond Davis Jr (1914–2006) in front of a prototype neutrino detector.**

he paid up on his 25-year-old bet by sending them a case of champagne. In 1995 Reines received a share of the Nobel Prize for this work; but Cowan had died in 1974 so could not receive this honour.

This was not the end of the story. Because neutrinos are so reluctant to interact with anything, the neutrinos from the heart of the Sun escape into space and cross past the Earth, and through it, almost unnoticed. Tens of billions of them pass through every square centimetre of your skin every second. Astronomers realized that if some of these solar neutrinos could be detected and analysed, they would provide a direct insight into what was going on in the very centre of the Sun.

It was another 'impossible' experiment, but Raymond Davis, of the Brookhaven National Laboratory, decided to give it a try. His detector had to be shielded from all sources of interference, such as cosmic rays (particles from space), so it was constructed 1,500 metres below ground at the bottom of the Homestake gold mine in Lead, South Dakota. Seven-thousand tons of rock had to be removed to make room for the detector, a tank the size of an Olympic swimming pool, filled with 400,000 litres of perchlorethylene, a fluid that used to be used in 'dry cleaning' processes.

The key component of this fluid was chlorine. On the rare occasions that a solar neutrino interacted with a chlorine atom in one of the molecules of perchlorethylene, it would convert the chlorine atom into an atom of a radioactive form of argon, which would be released into the liquid. Every few weeks, the fluid in the tank had to be swept clean of argon, by bubbling helium through it, and the number of argon atoms counted by detecting their radioactive decays. The experiment began operating in 1968. After all that effort, each run of the experiment yielded about a dozen counts. One radioactive argon atom was being produced in the tank every few days.

Once Davis had proved that solar neutrinos could be detected, other experiments were devised, and 'neutrino astrophysics' is now an important branch of astronomy which has provided insights into the workings of the Sun and into the nature of the neutrinos themselves. Davis received a share of the Nobel Prize, in 2002, for 'pioneering contributions to astrophysics, in particular for the detection of cosmic neutrinos'.

Nᵒ· 88   A VITAL VITAMIN

The application of X-ray crystallography in experiments to determine the structure of biological molecules, in particular proteins, was developed by Dorothy Crowfoot Hodgkin, following her early work with J. D. Bernal (see page 195). She made so many contributions that it is hard to pick out one

**Dorothy Crowfoot Hodgkin
(1910–1994).**

experiment as more important than the others – indeed, the Nobel committee didn't try, awarding her the Nobel Prize in Chemistry in 1964 'for her determinations by X-ray techniques of the structures of important biochemical substances'. This made her only the third woman to receive the chemistry prize, following Marie Curie and her daughter Irène. As the citation suggests, what really mattered was the experimental technique that Hodgkin applied to all the problems she tackled.

A key step in the analysis of the X-ray diffraction patterns, which were recorded photographically, was to use chemical techniques to insert heavy atoms at known locations in the molecules of the crystals being investigated. These extra atoms produce a distinctive 'signature' in the X-ray patterns, and by mapping out their locations it is possible to get an insight into the overall structure of the crystal. This is a long and tedious process. The photographs reveal the way electrons are distributed in the crystal, but the pattern for the whole crystal is too complicated to be worked out in one go. One region has to be analysed first, to provide a local map of the electron density, then this information helps to map out a neighbouring region, and so on. This involved a lot of mathematical analysis, initially without the aid of electronic computers.

In the 1940s, Hodgkin was working in Oxford, where she learned about the breakthrough research on penicillin from Ernst Chain (see page 189). By 1943, chemical analysis had established that the penicillin molecule contains 11 hydrogen atoms, 9 carbon atoms, 4 oxygen atoms, 2 nitrogen atoms, and 1 sulphur atom. But this combination of atoms could be arranged in either of two different ways, and it was essential to know which structure was correct if

Molecular model of
vitamin B12 (cobalamin).
The chemical formula is
$C_{63}H_{88}CoN_{14}O_{14}P$.

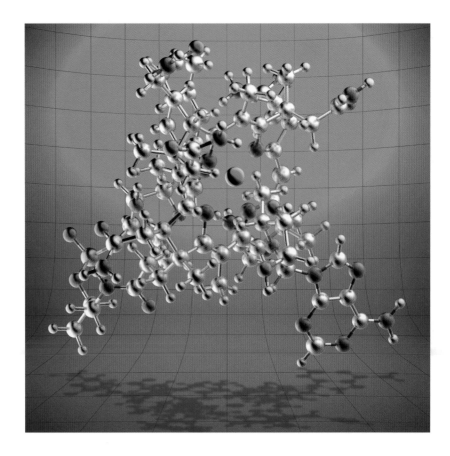

penicillin was to be manufactured in large quantities and similar antibiotics were
to be synthesized. Hodgkin took on the task of finding out.

Fortunately, the single sulphur atom in these molecules was heavy enough to
provide key information about the structure without the need to introduce
another heavy atom, and, after a great deal of analysis, Hodgkin was able to
work out which of the structures corresponded to biologically active penicillin.
After she had worked this out, the results were checked using a machine which
is sometimes described as a computer but was really a glorified calculator, not a
proper Turing machine (see page 209). The results were published in 1945.

Hodgkin then turned her attention to vitamin B12. This had recently been
identified as an essential substance used by the body to make red blood cells. A
lack of vitamin B12 causes a debilitating illness known as pernicious anaemia.
Vitamin B12 is present in foods derived from animals but not in vegetables, so
vegans, who do not eat dairy products, meat, fish, or eggs, may need supplements
of the vitamin. Analysing its structure, though, proved much harder than
analysing penicillin, because it is a much larger and more complicated molecule.
Although Hodgkin started work on the project in 1948, it was only completed in
1956, with the results being published a year later, and represents her greatest

achievement as a crystallographer. This time, however, she did have the help of a genuine computer, although it was not in Oxford, or even in England. She collaborated with Kenneth Trueblood, the head of a team of American crystallographers based at the University of California in Los Angeles, who had access to a computer. Hodgkin provided the data obtained by crystallography for analysis, and the UCLA team ran the calculations. Long before the advent of email and the Internet, they exchanged the information by post.

In her Nobel lecture, after describing her successes up to that point, Hodgkin said: 'I should not like to leave an impression that all structural problems can be settled by X-ray analysis or that all crystal structures are easy to solve. I seem to have spent much more of my life not solving structures than solving them. I will illustrate some of the difficulties to be overcome by considering our efforts to achieve the X-ray analysis of insulin.'[47]

She then described how she had been tackling the analysis of insulin, the drug used to treat diabetes, off and on since the mid1930s, with only limited success. The molecule contains 777 atoms (penicillin has just 27 atoms), and has an extremely complex structure. But Hodgkin was still not finished with insulin, and, aided by improved computer technology, she did solve the structure. This work was completed in 1969, five years after she received her Nobel Prize.

Nᵒ. 89   THE BREATHING PLANET

One of the most important experiments of the twentieth century began in the late 1950s and is still running. It shows how our entire planet 'breathes' with the seasons, and revealed the build up of carbon dioxide in the atmosphere that threatens a potentially catastrophic global warming over the next hundred years. But it was another serendipitous discovery.

In the early 1950s, Charles David Keeling, a young researcher at the California Institute of Technology (Caltech), planned to study the balance of carbon dioxide among the air, the oceans, and in limestone rock. As a step towards this, he developed a very precise instrument, called a gas manometer, to measure the concentration of carbon dioxide in the air. Because the air in Pasadena, where Caltech is located, was polluted by human activities, he tested the instrument at Big Sur, near Monterey, where he found that the air contained slightly more carbon dioxide at night than in the day time, because of the respiration of plants. But the average was the same every afternoon – about 310 parts per million (ppm).

Measurements in other American forests showed the same pattern. Both meteorologists and oceanographers were intrigued by the observations, and in 1956 Keeling, by then based at Scripps Institution of Oceanography, obtained funding for further research as part of the International Geophysical Year. This was a worldwide programme of research into many aspects of planetary science, which officially ran from 1956 to 1957 but initiated many longer-term experiments. Keeling's plan was to measure, with the aid of colleagues, atmospheric carbon dioxide concentrations at several locations around the globe, including on top of Mauna Loa in Hawaii. This volcanic mountain is in the middle of the Pacific Ocean, far from any forests or sources of industrial pollution. He began operating the Mauna Loa experiment himself, in March 1958, recording a concentration of 313 ppm on the first day.

To Keeling's surprise, the concentration of carbon dioxide rose slightly through March, April and May, then declined until October, when it began to rise again until the following May, when the pattern was repeated. Indeed, it repeated every year. The cycle showed that the whole planet was breathing, driven by the fact that there is more land, and therefore more land-based vegetation, in the northern hemisphere than in the south. In northern hemisphere in summer, this vegetation grows, taking up carbon dioxide from the air. In winter, it dies back and decays, releasing carbon dioxide. This produces a cyclic variation of about 5 ppm each year.

After the seasonal effects were allowed for, however, the average concentration of carbon dioxide in the air was higher in 1959 than in 1958, and higher still in 1960. Keeling published scientific papers reporting this, regularly updating what became known as the 'Keeling curve', a graph showing the build up of carbon dioxide in the atmosphere with the seasonal cycles superimposed onto it. By the 1970s, with growing concern that this build up of carbon dioxide might contribute to global warming through the so-called greenhouse effect, it was becoming clear that the increase was due to human

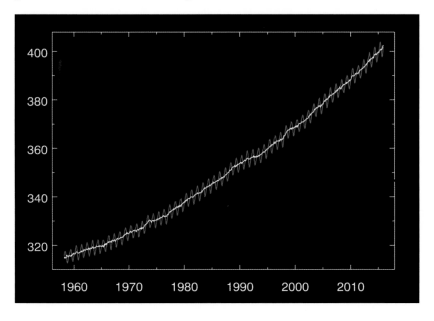

The 'Keeling Curve' – Mauna Loa CO_2 record, 1958–2015.

Charles Keeling (1928–2005) photographed in March 1992 in his laboratory at the Scripps Institution of Oceanography, San Diego, USA.

activities, including forest clearance and the burning of fossil fuels. This was confirmed in the 1970s, when a political crisis caused a large increase in crude oil prices and a consequent slowdown in burning oil-based fuel. The rise of the Keeling curve levelled off slightly to match the change in these anthropogenic sources of carbon dioxide.

By 2015, the concentration of carbon dioxide in the air had risen above 400 ppm, which represents an increase of roughly 28 per cent above the level in 1958. Comparison with the amount of fossil fuel being burnt around the world shows that 57 per cent of the carbon dioxide produced remains in the air, while the rest is absorbed by poorly understood natural 'sinks', including the oceans. In the longer-term context, studies of air bubbles trapped in ice cores from polar regions show that over the past 400,000 years, carbon dioxide concentrations were around 200 ppm during ice ages, and around 280 ppm during the warmer intervals known as interglacials. The concentration started rising further at the beginning of the nineteenth century. This suggests that even in 1958 the concentration was above the 'natural' level, thanks to human activities over the previous one and a half centuries. As carbon dioxide is known to trap heat near the surface of the Earth (see page 110), it is clear that this build up of carbon dioxide has contributed to, and is most probably the principal cause of, the global warming that the Earth has experienced since at least the middle of the twentieth century. Because of the importance of these implications, carbon-dioxide concentrations are now monitored at about a hundred sites around the world, but the Mauna Loa record is the longest and still has pride of place. These observations are now supervised by Keeling's son Ralph, a professor at Scripps.

In 1963, two young researchers, Arno Penzias and Robert Wilson, joined forces to adapt a radio telescope, originally designed to establish the feasibility of global satellite communications, for astronomical work. The telescope, at Crawford Hill in New Jersey, was owned by the Bell telephone company, which had a policy of allowing its research teams time for purely scientific projects. Before this team could start making astronomical observations, they had to calibrate the telescope and remove all sources of interference – 'noise'. It was during this process that they unexpectedly made a discovery that would earn them the Nobel Prize.

The business end of the telescope incorporated a very sensitive receiver to detect weak radio emissions from objects in space. Astronomers call these 'signals', but they are not artificial signals like TV transmissions, just the natural radio emissions produced by astronomical objects such as active galaxies. The strength of these signals is measured in terms of the temperature of equivalent blackbody radiation, and the receiver, a radiometer using maser amplifiers, was so sensitive that it could detect radiation as cold as a few degrees on the Kelvin scale (K), just above *minus* 273 degrees Celsius.

To calibrate the telescope, it was pointed at the sky (at nothing in particular) and the radiometer was switched between the signal coming from the antenna and the signal from a 'cold load' kept at a temperature of just over 4 K using liquid helium. This showed them how much hotter or colder the antenna signal was than the signal from the cold load. Penzias and Wilson expected the 'temperature of the sky' to be zero, so in this way they could make a calibration and subtract out all the known sources of noise, such as the temperature of the air above the antenna. What was left would be noise due to the antenna itself, which they intended to remove by appropriate techniques, such as polishing the metal to a smooth surface. But no matter how they tried, the team were never able to reduce the noise in the system to zero. They even tried laboriously cleaning out the droppings left in the antenna by nesting pigeons, and sticking shiny aluminium tape over all the riveted joints, but to no avail. They were left with what they called an 'excess antenna temperature' of between 2 K and 3 K, meaning that the radiation coming in to the radiometer from the antenna was at least 2 K hotter than they could explain. The signal was the same day and night, day after day and week after week. It corresponds to a very weak hiss of radio noise at microwave frequencies, like the 'signal' from a very cold microwave oven.

Throughout 1964, Penzias and Wilson remained baffled, and their whole radio astronomy project hung in the balance. But then news of their predicament reached a team of radio astronomers at nearby Princeton

Holmdel horn antenna, Bell Laboratories, New Jersey, USA.

University. They were interested in the possibility that the Universe as we know it had emerged from a hot, dense state (the Big Bang), and they had calculated that this event should have filled the Universe with electromagnetic radiation that would by now have cooled to a temperature of a few K (unknown to them, the same prediction had been made in the 1940s, by Ralph Alpher and Robert Herman, but had been ignored). The Princeton team were in the process of building a radio telescope to look for this radiation when news of the work at Crawford Hill reached them.

When the two teams got together to discuss the discovery, it became clear that the radio noise from space found by Penzias and Wilson might be the echo of the Big Bang. Penzias and Wilson were not entirely convinced, but they were relieved to have any explanation, and felt confident enough to publish their results in 1965, in a paper with the cautious title 'A Measurement of Excess Antenna Temperature at 4080 Mc/s', without actually claiming that they were detecting radiation from the birth of the Universe. As Wilson put it in his Nobel lecture: 'Arno and I were careful to exclude any discussion of the cosmological theory of the origin of background radiation from our letter because we had not been involved in any of that work. We thought, furthermore, that our measurement was independent of the theory and might outlive it.'[48]

But the announcement triggered a wave of further experiments which confirmed that the Universe is filled with a sea of microwave radiation at a temperature of 2.712 K, and this remains compelling evidence that there was a Big Bang. In the twenty-first century, observations of this microwave background radiation have refined our understanding of the Universe and revealed details of its properties (see page 274).

In 1978, Penzias and Wilson received a share of the Nobel Prize in Physics 'for their discovery of cosmic microwave background radiation'.

Nᴼ· 91   CLOCKING ON TO RELATIVITY

It is a well-known prediction of the special theory of relativity that a clock moving past an observer will be seen to run slow compared with the observer's clocks – time dilation. It is a less well-known prediction of the general theory of relativity that a clock in a gravitational field will be seen to run slow compared with the clocks of an observer in a weaker gravitational field. This is linked to an observed change in the wavelength of light coming from a source in a gravitational field, which is known as the gravitational redshift. These predictions were tested in 1971 by an experiment that involved flying clocks around the world in ordinary commercial aircraft.

The experiment was carried out by two American physicists, Joseph Hafele and Richard Keating. The idea was to see how the time recorded on moving clocks at high altitude (combining the effects of both the special theory and the general theory) compared with the time recorded by clocks that stayed at rest on the ground. It would have been a lot easier for them if they could have chartered a private plane, but because of a limited budget they had to fly economy class on ordinary scheduled flights. The ultra-precise atomic clocks that they used were strapped to the front wall of the passenger cabin and connected to the power supply of the aircraft; an identical set of clocks stayed at the US Naval Observatory in Washington, DC, ready to be compared with the travelling clocks when they returned home.

Between 4 and 7 October 1971, the clocks were flown completely around the world from west to east. After making allowances for the various stopovers, changes of altitude and changes of speed made by the aircraft, the team calculated that the clocks should have gained between 254 and 296 nanoseconds (billionths of a second), with two-thirds of this attributable to the gravitational effect of being at altitude (the clocks ran faster, because gravity is weaker at altitude). The rest of the calculated difference was due to the effect predicted by the special theory, which added to the gravitational effect because of complications caused by the Earth's rotation. The simplest way to picture this is to think in terms of a 'frame of reference' stationary relative to the centre of the Earth. A clock in an aircraft moving eastward, in the direction of the Earth's rotation, moves faster than one on the ground, but a clock aboard an aircraft moving westward moves slower than one on the ground. The measured difference was 273 nanoseconds, smack in the middle of the predicted range.

The same clocks were then flown westward around the world, between 13 and 17 October 1971. The results from this leg of the experiment were slightly less impressive. This time, the time-dilation effect from the motion of the clocks was expected to cause the clocks to lose more than they would gain from the gravitational effect, producing a time difference of 40 ± 23 nanoseconds compared with the clocks on the ground. Because of some problems with the data measurements, this time the team could say only that the measured difference was between 49 and 69 nanoseconds, which just about matched the prediction. But together the two legs of the experiment confirmed, if anyone had doubted it, that the time-dilation effects predicted by Einstein's two great theories were real.

A more accurate test along the same lines was carried out in June 1976, involving the Smithsonian Astrophysical Observatory and NASA. A rocket-borne experiment known as Gravity Probe A reached an altitude of

Joseph Hafele and Richard Keating with their atomic clock.

10,000 kilometres (far higher than any aircraft) on a flight going almost straight up and down that lasted just under two hours. This meant that effects due to the Earth's rotation could be ignored. At the top of its flight, the payload felt a gravitational influence only 10 per cent as strong as the influence we feel on the surface of the Earth.

During the flight, the time recorded on a clock in the payload of the rocket, incorporating a maser, was monitored over a radio link and compared with the timekeeping of an identical clock on the ground while the flight was still in progress. The changing speed of the rocket and its payload was also monitored, by the Doppler effect (see page 109), using a second signal transmitted from the Earth and received and retransmitted by the probe back to Earth. The observations were then compared with the predictions calculated using the equations of relativity theory. This time, the measurements matched the predictions to a precision of 70 parts per million, or seven thousandths of 1 per cent. Nobody was surprised, but a lot of physicists were very pleased; this is still the most complete and accurate measurement of the gravitational redshift yet carried out.

N°. 92   MAKING WAVES IN THE UNIVERSE

One of nature's most spectacular experiments involves a pair of stars, each about the size of a large mountain on Earth but with more mass than the Sun, locked in a gravitational embrace, orbiting around each other at speeds of hundreds of kilometres per second. Observations of this system, dubbed the binary pulsar, show that it is producing ripples in the fabric of space itself – gravitational waves.

This kind of gravitational radiation was predicted as long ago as 1916, by Albert Einstein. But the discovery confirming the accuracy of his prediction was not made until 1974. That year, Russell Hulse, a young astronomer working with the Arecibo radio telescope in Puerto Rico, noticed something odd about the behaviour of a radio star known as a pulsar. Pulsars are fast-spinning neutron stars, tiny but very dense objects (as dense as the nucleus of an atom), which emit beams of radio noise that sweep around like the beam of a lighthouse. A neutron star has a radius of about 10 kilometres, and the force of gravity at the surface of such a star is a hundred-thousand-million times more than on Earth. If the Earth happens to be in the path of the beam from a pulsar, radio telescopes pick up a regular pulse of radio waves, like the ticking of a clock. This particular pulsar, known as PSR 1913+16, spins on its axis once every 0.059 seconds, making it one of the fastest pulsars known.

Most pulsars are superbly accurate clocks, beating time with a precision measured to many decimal places. But during a series of observations of this pulsar in the summer of 1974 Hulse found that it has a period which changes by as much as 30 microseconds from one day to the next – a huge 'error' for a pulsar. This variation follows a rhythm of its own, changing the measured period over a regular cycle. He realized that this could be a result of the changing Doppler effect (see page 106) caused by the pulsar moving in a tight orbit around a similar star that was not emitting any detectable radio noise.

Hulse's colleague Joseph Taylor (they both worked at the University of Massachusetts) joined Hulse in Arecibo to carry out a more detailed investigation. Together, they found that the pulsar zips round its companion once every 7 hours and 45 minutes, reaching a maximum speed of 300 kilometres per second and with an average speed of about 200 kilometres per second. The size of the orbit is about 6 million kilometres, about the same as the circumference of the Sun. So the whole

An artist's depiction of the Arecibo radio telescope.

A hypothetical binary pulsar. One neutron star (lower centre) is emitting a pulse of energy. Pulsars are rapidly rotating neutron stars that cast out narrow beams of energy as they rotate. The dark pink ellipses are the mutual orbits of these pulsars around their mutual centre of mass (blue dot). The orbits are slowly moving outward as the system loses energy by gravitational radiation. (The effect is hugely exaggerated here!)

binary system would fit inside the Sun. The orbital properties also told them that the combined mass of both stars added up to 2.8275 times the mass of the Sun.

Astronomers immediately realized that such an extreme system would provide a test bed for Einstein's ideas about gravitational radiation, which were based on his general theory of relativity. According to the general theory, under these extreme conditions the orbiting stars should be producing ripples in space, like the ripples you might imagine being produced in a tank of water by a rotating dumbbell. This gravitational radiation would carry energy away from the system, altering the orbit. It would cause the orbital period (which in round numbers is 27,000 seconds) to increase by 75 millionths of a second per year – about 0.0000003 per cent per year. In 1978, after four years of observations, this change had been measured accurately enough to confirm that Einstein was right, that gravitational radiation really exists. By 1983 the change had been measured to an accuracy of 2 millionths of a second per year, giving a value of 76 ± 2 millionths of a second per year. Since then, the accuracy of the general theory has been confirmed to better than 1 per cent.

The extended observations also made it possible to work out the ratio of the masses of the two stars, partly from the time-dilation effect of the pulsar's high speed on its timekeeping. Along the way, this confirmed the accuracy of the special theory of relativity. As they already knew the total mass, this ratio enabled Hulse and Taylor to work out that the mass of PSR 1913+16 itself is 1.42 times the mass of the Sun, while its companion has a mass of 1.40 times that of the Sun. These were the first accurate measurements of the masses of neutron stars.

Since the discovery of the binary nature of PSR 1913+16, other similar systems have been discovered, always confirming the accuracy of the general theory. But to astronomers, PSR 1913+16 is still known as 'the' binary pulsar. In 1993, Hulse and Taylor shared the Nobel Prize in Physics 'for the discovery of a new type of pulsar, a discovery that has opened up new possibilities for the study of gravitation.'

Nº 93   THE PACEMAKER OF ICE AGES

Why does the Earth experience Ice Ages? It seems a natural question, but it turns out to have been phrased the wrong way round. The real puzzle is why, given the present-day arrangement of the continents, the Earth is not in a permanent Ice Age. The answer was suggested more than a hundred years ago, but confirmed by experiment only in the mid-1970s.

It was only in the nineteenth century that geologists realized that scratches and scars in the rocks, and other evidence, show that great ice sheets had once (perhaps more than once) spread southwards across Europe and North America from the polar regions (see page 108). The Scot James Croll, in the nineteenth century, and the Serbian Milutin Milankovitch, in the first half of the twentieth century, suggested that this might be related to the way the Earth tilts and wobbles as it orbits the Sun; but the idea, which became known as the Milankovitch Model, was not widely accepted.

We experience seasons because the spinning Earth leans over at an angle of roughly 23.5 degrees from the vertical as it orbits around the Sun. This means that first one hemisphere leans towards the Sun, bringing summer, while at the same time, the other hemisphere experiences winter. Six months later the situation is reversed. But this tilt is not constant. There is a wobble caused by the gravitational influence of the Sun and Moon, so that it changes from 21.8 degrees to 24.4 degrees over thousands of years. And there are other, more subtle, orbital variations. All this changes the amount of heat received from the Sun at different

Milankovitch cycles. The Earth's axis completes one full cycle of precession approximately every 26,000 years (top diagrams). At the same time, the orbit's eccentricity varies with a period of about 400,000 years (bottom right). In addition, the angular tilt of Earth's rotational axis varies from 22.1 degrees to 24.5 degrees and back again on a 41,000-year cycle. Currently, this angle is 23.44 degrees and is decreasing.

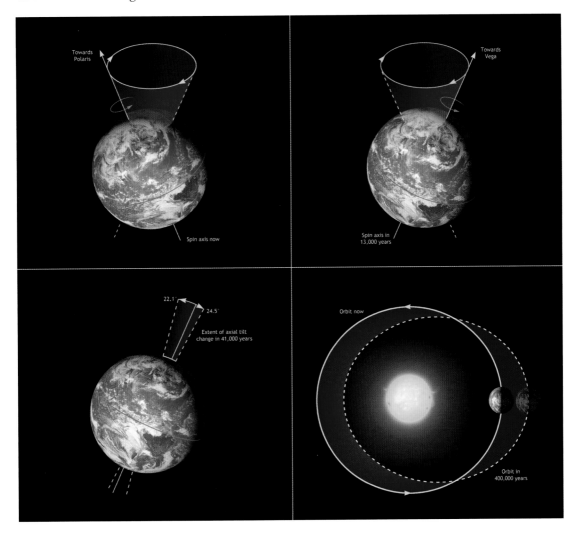

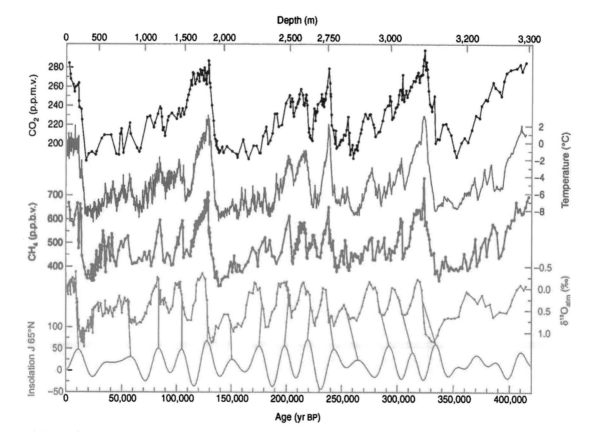

Depth (m)

Age (yr BP)

Ice age rhythms. 420,000 years of ice core data from the Vostok Antarctica research station reveal changes in the concentration of isotopes of oxygen, the concentration of methane, and the amount of carbon dioxide in the air. The oxygen isotopes reveal changes in temperature. All this is compared with the changing insolation at 65 degrees N due to the Milankovitch cycles, which matches the oxygen/temperature curve. Notice the sudden jump in the upper curves in recent decades (left), going out of step with the astronomical cycles. This is a result of human activities that are warming the globe.

latitudes (the 'insolation') in different seasons, although the amount of heat received by the whole planet over a whole year stays the same.

Milankovitch calculated just how all these changes affected the insolation over hundreds of thousands of years – a remarkable tour de force in the days before electronic computers. It was natural to expect that Ice Ages would be at their peak when northern hemisphere winters were coldest, so ice could spread over the land. As Antarctica is always covered in ice and there is no nearby land to be affected, the southern hemisphere does not have such a powerful potential influence. But as geologists gathered more data they found exactly the opposite of the anticipated pattern. At times when Milankovitch calculated that northern winters were coldest, Ice Ages went in to retreat. The pattern that began to emerge was one in which, for several million years, during the Quaternary Period of geological time, the Earth has been in the grip of an Ice 'Epoch', with glaciers spreading in full Ice Age conditions for roughly a hundred-thousand years then retreating temporarily during intervals of roughly 10,000 years, known as interglacials. We are living in an interglacial.

Then, theorists realized what is going on. Because the heat balance over a year is the same, very cold northern winters go hand in hand with very hot northern summers. Interglacials occur only when the northern hemisphere summers are hot enough to melt the ice back towards the Arctic.

This was fine in principle, but how could it be tested? The key experiment depended on the discovery that the Earth's magnetic field changes in strength and sometimes reverses direction, leaving a trace of fossil magnetism in the rocks (see page 232). Using this fossil magnetism, and evidence from other tracers, geologists were able to date the layers of sediment in cores drilled from the bed of the southern ocean, centrally located between Africa, Australia, and Antarctica. These sediments are laid down as the years pass, with the youngest at the top, and remain undisturbed on the sea bed. By the mid1970s, geologists had a continuous record going back half a million years. These cores contain the remains of tiny sea creatures known as foraminifera, and the shells of these creatures contain atoms of oxygen that came from the water in which those creatures lived, which they used in building those shells. Oxygen in sea water comes in two varieties (isotopes), oxygen-16 and the heavier oxygen-18. Heavier molecules freeze more easily than light molecules, so during an Ice Age there are proportionally fewer molecules of water containing oxygen-18 around for the foraminifera to absorb. So by measuring the proportions of oxygen-16 and oxygen-18 in the organic remains from the deep sea sediments, climatologists were able to infer how the ice sheets had advanced and retreated over the millennia, with enough precision to test the predictions of the revised Milankovitch Model.

In 1976, Jim Hays, John Imbrie, and Nick Shackleton published a paper in *Science*, with the title 'Variations in the Earth's Orbit: Pacemaker of the Ice Ages', which pulled together all of this evidence and confirmed that interglacials occurred only when northern hemisphere summers were at their hottest. 'It is concluded,' they said, 'that changes in the earth's orbital geometry are the fundamental cause of the succession of Quaternary ice ages.'[49] It was the moment when the Milankovitch Model came in from the cold and, having passed the experimental test, became a theory.

No. 94   THE WORLD IS NON-LOCAL

Common sense tells us that if I hit a cricket ball on a playing field in England, this has no effect on a cricket ball in Australia, even if the two balls were manufactured in the same batch in the same factory and once nestled together in the same box. But does the same common sense apply to things in the quantum world, such as photons and electrons? Bizarre though it may seem, in the twentieth century, quantum physics raised the real possibility that the answer might be 'no'. This was eventually proved in an experiment carried out in 1982.

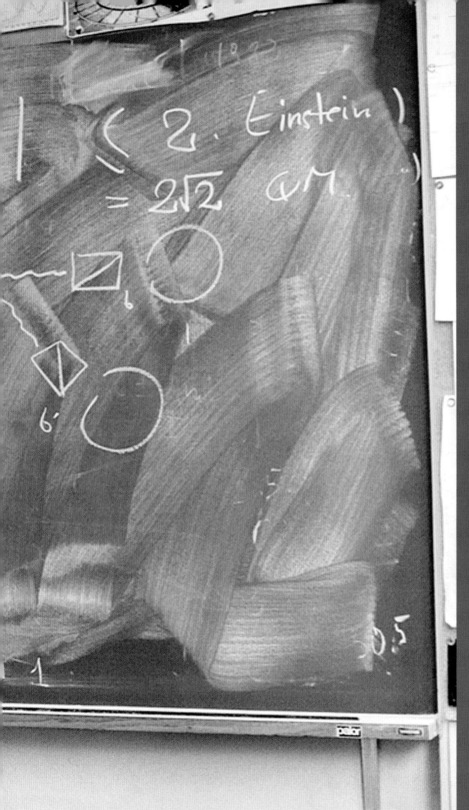

John Bell (1928–1990).

It all started in 1935, when Albert Einstein and his colleagues Boris Podolsky and Nathan Rosen presented a puzzle (sometimes known as the 'EPR Paradox') in the form of a thought experiment. This was later refined by David Bohm, and later still by John Bell. In its later form, the puzzle concerns the behaviour of two photons (particles of light) ejected from an atom in opposite directions. The photons have a property called polarization, which can be thought of as like carrying a spear pointing either up, down, or at any angle across the direction of travel. The key feature of the puzzle is that the photons must have different polarization, but correlated in a certain way. For simplicity, imagine that if one photon is vertically polarized the other must be horizontally polarized.

Now comes the twist. Quantum physics tells us that the polarization of the photon is not determined – it does not become 'real' – until it is measured. The act of measurement forces it to 'choose' a particular polarization, and it is possible (indeed, straightforward) to set up an experiment which forces a photon to be vertically polarized, or forces it to be horizontally polarized, whichever you wish. This is scarcely any more sophisticated than letting the light shine through a lens of a pair of polarizing sunglasses. The essence of the EPR 'paradox' is that, according to all this, measuring one of the pair of photons and forcing it to become, say, vertically polarized instantaneously forces the other photon, far away and untouched, to become horizontally polarized. Einstein and his colleagues said that this is ridiculous, defying common sense, so quantum mechanics must be wrong.

After John Bell presented the puzzle in a particularly clear form in the 1960s, the challenge of testing the prediction was taken up by several teams of experimenters, leading up to a comprehensive and complete experiment carried out by Alain Aspect and his colleagues in Paris in the early 1980s. Although such experiments have since been refined and improved, they always give the same results that emerged from the Aspect experiment itself.

The key feature of the experiment is that the choice of which polarization will be measured is made automatically and at random by the experiment, after the photons have left the atom. At the time the 'forced' photon arrives at the polarizer, there has not been long enough for any signal, even travelling at the speed of light, to have reached the other side of the experiment. So there is no way that the detector used to measure the second photon 'knows' what the first measurement is.

It would be very difficult (just about impossible, even with present technology) to do the experiment literally with pairs of photons, two at a time; but in the Aspect experiment and its successors very many pairs of photons are studied, with more than two angles of polarization being investigated, and the results analysed statistically. John Bell's great contribution was to show that in this kind of analysis if one particular number that emerges from the statistics is bigger than another specific number, common sense prevails and there is no trace of what Einstein used to call 'spooky action at a distance'. This is what Bell expected to happen, and it is known as Bell's Inequality. But the experiments show that Bell's Inequality is violated. The first number is smaller than the second number. Experiments are somehow particularly convincing when they prove the opposite of what the experimenters set out to find – it certainly shows that they were not cheating, or unconsciously biased by their preconceptions! But what does it mean?

The pairs of photons really are linked by spooky action at a distance, confounding 'common sense', in a state that quantum physicists call entanglement. What happens to photon A really does affect photon B, instantaneously, no matter how far apart they are. This is called 'non-locality', because the effect is not 'local' (specifically, it occurs faster than light, although it turns out that no useful information, such as the result of the 3.30 race at Newmarket, can be transmitted faster than light by this or any other means). The Aspect experiment and its successors show that the world is non-local. And this strange property even has practical implications, as in the world of quantum computing (see page 268).

THE ULTIMATE QUANTUM EXPERIMENT

 definitive version of the experiment which demonstrates what Richard Feynman called 'the central mystery' of quantum physics was carried out by a Japanese team at the end of the 1980s. This was at the time the ultimate version of the double-slit experiment (see page 79), using individual electrons to demonstrate both wave-particle duality and the holistic nature of the quantum world.

In the classic version of the double-slit experiment, light is sent through two holes and spreads out on the other side to make an interference pattern, proving that light is a wave. In the version developed by Akira Tonomura and his colleagues at the Hitachi research laboratories, individual electrons were fired one at a time past a very thin conducting wire at right angles to the path of the electrons to give them a choice of two routes past the wire. This setup is known as an electron biprism. On the other side of the wire there was a screen, essentially the same as a TV screen or a computer screen, on which each electron made a spot of light as it arrived. But the spots did not just flash on and off; each spot stayed on the screen as more electrons arrived. And the whole thing was recorded, so that the team had a movie showing the pattern that was building up on the screen as more and more electrons arrived and more spots were produced.

If electrons behaved like particles in the everyday world – like tennis balls, perhaps – you would not expect much of a pattern to build up at all. The balls

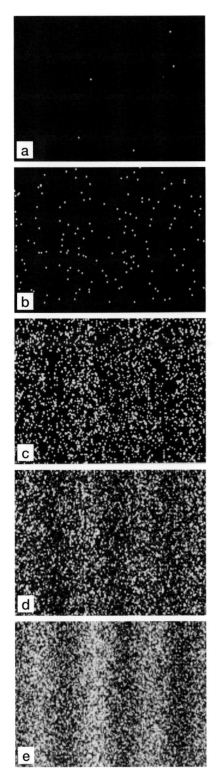

that went one way round the wire would combine to make a blob on one side of the screen, and the balls that went the other way round the wire would combine to make a blob on the other side of the screen. But that is not what the Hitachi team saw. Each electron did indeed make a single spot on the screen, and at first these spots seemed to be distributed randomly across the screen. But as more spots arrived, they made a pattern. The pattern that built up was the typical stripy pattern produced by interference. Although the electrons started out as particles and arrived as particles, on their way through the experiment they seem to have behaved like waves, apparently interfering with each other even though only one electron was passing through the biprism at any time. They seemed to be 'aware' of the whole experimental setup, in both time and space.

The Hitachi team, who published their results in 1989, were not the only people to observe this phenomenon. Pier Giorgio Merli, Giulio Pozzi, and Gianfranco Missiroli carried out interference experiments with single electrons in Bologna in the 1970s. There has been some debate about who deserves priority, but the Japanese team seems to have dotted the 'i's and crossed the 't's. Pozzi and his colleagues also carried out the first such experiment using an actual double slit and announced their results in 2008. As expected, they showed interference.

This work helped to encourage others to go a step further and build an experiment which not only literally uses two slits, instead of an electron biprism, to make electrons interfere, but in which the opening and closing of the slits can be controlled at will. After three years of work, a team headed by Herman Batelaan, of the University of Nebraska-Lincoln, announced their results in 2013. In their version of the experiment, electrons were fired one at a time at a wall made of a gold-coated silicon membrane. There were two slits in the wall, each 62 nanometres wide, with the centres of the slits 272 nanometres apart. One or both slits could be closed by a tiny sliding shutter whenever the experimenters chose. As usual, the electrons were captured on a screen after they had passed through the slits. Only about one electron was detected each second.

With just one slit open, they observed the build up of the same kind of blob that you would expect if electrons behaved like tennis balls. But with both slits open they saw an interference pattern, just as the Italian and Japanese teams had seen in their experiments. The electrons 'knew' how many slits were open.

Don't worry if you cannot understand how this is possible. As Feynman said in his book, *The Character of Physical Law*, 'I think I can safely say that nobody understands quantum mechanics.' And he advised, 'Do not keep saying to yourself, if you can possibly avoid it, "But how can it be like that?" because you will get "down the drain", into a blind alley from which nobody has yet escaped. Nobody knows how it can be like that.'

And yet, in spite of not understanding how quantum physics can be like that, scientists are able to put it to practical use, not least in the realm of quantum computing (see page 268).

Nº. 96 THE ACCELERATING UNIVERSE

In the 1990s, two teams were mapping the expanding Universe by studying light from the exploding stars known as supernovae, observed in distant galaxies. A particular class of supernovae, part of a family known as SN 1a, are very useful for this work because they all reach the same maximum brightness during the course of the explosion. So if astronomers can measure the brightness of a distant supernova, they can work out how far away the galaxy it resides in is, in the same way that measurements of the brightness of Cepheids established that spiral nebulae are galaxies beyond the Milky Way and that the Universe is expanding (see page 170).

This was very difficult work, pushing the technology to the limit. On average, about two supernovae go off in a galaxy every thousand years, so you need to look at 50,000 galaxies to detect about a hundred supernovae each year. By repeatedly photographing dozens of patches of sky, each containing hundreds of very faint galaxies, the teams hoped to find a handful of supernovae each year. By comparing the peak brightness of these supernovae with the redshifts of their host galaxies, they intended to measure the rate at which the expansion of the Universe is slowing down as gravity tries to halt the expansion that started with the Big Bang. Because light takes a finite time to travel across the Universe, by looking at distant galaxies astronomers are looking back into the past, seeing how the Universe has changed as time passes.

To their surprise, the two teams independently discovered something odd. For very faint galaxies (that is, very distant galaxies), the distances calculated from the observations did not quite match their expectations. The supernovae were a little bit too faint, according to the standard redshift-distance rule. This meant one of two things. Either the distant supernovae really were fainter than their nearer cousins, or they were further away than their redshifts implied. If they were further away, it meant that the Universe had expanded more than anticipated –

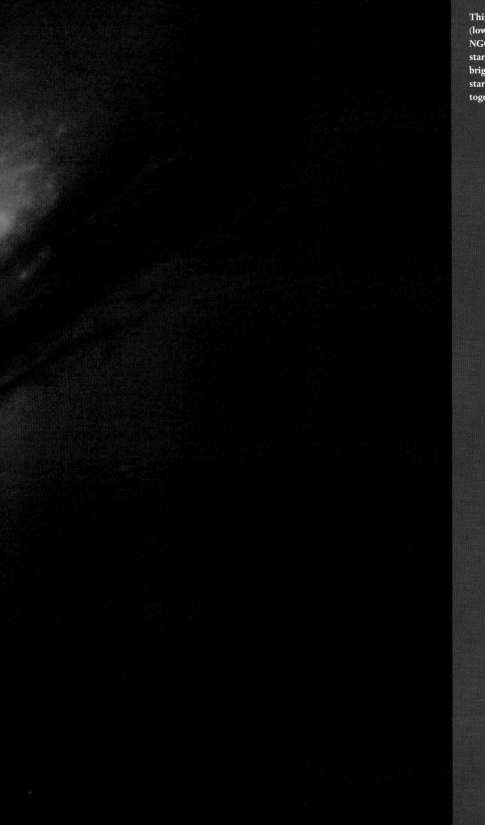

This image of a supernova
(lower left) in the galaxy
NGC 4526 shows a single
star briefly shining as
brightly as all the other
stars in its parent galaxy put
together.

Adam Riess (b. 1969).

that the expansion is speeding up, not slowing down. Something must be pushing against gravity.

There is no reason to think that the distant supernovae are any different from their nearby cousins, but although popular reports suggested that cosmology had been shaken to its foundations by the discovery, in fact cosmologists had a natural explanation for the source of the force that is making the Universe expand faster. When Albert Einstein used his general theory of relativity to develop equations that describe the nature of the Universe at large, he included a term known as the cosmological constant, which is a measure of the energy of empty space. This number had traditionally been set as zero, as it did not seem to be needed. But if it had a small value, that would mean that every cubic centimetre of space has the same amount of what has become known as dark energy. Dark energy contributes a kind of springiness to space, pushing outwards against the force of gravity.

Just after the Big Bang, although the Universe was expanding, gravity was slowing the expansion down. At that time, the outward urge of dark energy was too small to have much effect. But as the Universe expanded, there were more cubic centimetres of space, so there was more dark energy. At the same time, the gravitational pull of galaxies on each other got weaker as they moved apart. So there came a time when the outward urge from dark energy became bigger than the inward tug of gravity, and the expansion began to accelerate. The supernova results were announced in 1998, and quickly explained in this way in terms of dark energy. Since then, studies of even fainter supernovae – indicating more distant objects seen longer ago in time – have confirmed that when the Universe was younger the expansion was indeed slowing down.

One of the two teams who made the discovery was led by Brian Schmidt and Adam Riess, the other by Saul Perlmutter. In 2011 the three of them shared the Nobel Prize in Physics 'for the discovery of the accelerating expansion of the Universe through observations of distant supernovae'. As Perlmutter said in an interview with the BBC, 'The two groups announced their results within just weeks of each other and they agreed so closely; that's one of the things that made it possible for the scientific community to accept the result so quickly.'

But each team involved dozens of researchers working at various places around the world. It is now thought that dark energy makes up at least two-thirds of the mass-energy of the Universe (see page 274). Finding out more about dark energy and its implications for the fate of the Universe is now a key area of cosmological research.

MAPPING THE HUMAN GENOME

J ust 70 years ago – the span of a Biblical lifetime – the chemical composition of genes was still a matter of debate. Less than 60 years later, in 2001, the composition of every gene in the human body (the human genome) had been mapped. Armed with this information, it would be possible in principle, given the technology, to make a human being out of basic chemicals. More practically, it opens the way to cure diseases such as cystic fibrosis by genetic engineering, repairing faulty genes.

DNA, the stuff genes are made of, is composed of a string of bases known by the letters A, C, G and T (see page 172). These can be thought of as the letters in a four-letter alphabet. Instructions for the construction and operation of the body at a cellular level are coded into genes in the form of three-letter 'words', such as AGT, GAT, AAC, and so on. These are known as triplets. A string of these words contains information, just as a string of words in our

DNA sequencing of the human genome, being carried out automatically by a computer. Photographed in Leroy Hood's laboratory at Caltech.

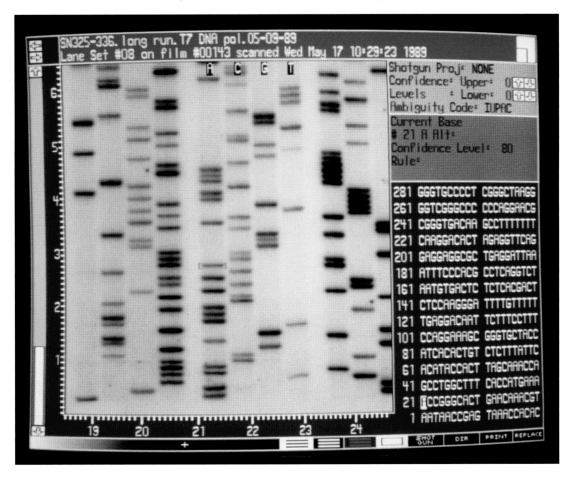

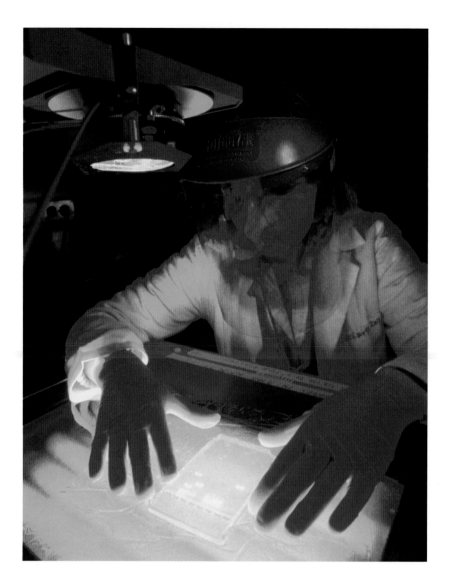

Scientist in a darkroom, preparing to photograph an electrophoresis gel used in mapping DNA extracted from chromosomes of the bacteria Escherichia coli.

alphabet contains information. Any information can be 'written' in this way if the string of letters is long enough – after all, any information that can be written in the English language can also be written in the binary code of computers, which contains only two letters, 0 and 1, on or off. That is how we are writing this book, and maybe how you are reading it, if you are using an e-reader.

The string of letters in DNA is long enough to contain the information required for human life. The human genome contains roughly 3 billion letters, equivalent to about a Gigabyte of computer memory. In a sense, this is surprisingly small, which is what made the project to map the human genome possible, but it applies to each individual cell. Thanks to gene mapping, we now know that there are between 20,000 and 25,000 separate human genes, arranged in 22 paired chromosomes plus the X and Y chromosomes that determine sex.

Mapping a gene (or a genome) is technically known as sequencing – finding the sequence of letters along a strand of DNA. The experimenters also need to find out where one gene ends and another begins along a strand of DNA that makes up a chromosome. To do all this they use enzymes, which act as chemical scissors, to cut DNA into small pieces at precise locations that depend on finding an exact string of letters, like finding a particular word in a book. These fragments are then separated by a technique called gel electrophoresis. The bits of DNA are put at one end of a tube filled with sticky gel, and an electric current is used to drag them through the gel. Small fragments encounter less drag, so they move faster than long fragments, and the bits get sorted by size. Then, the fragments are analysed chemically to find the sequence of bases along them. This involves 'amplifying' the fragments by making multiple copies of them using the same techniques that the body uses to copy DNA, so the chemists have a brew containing many identical copies of a stretch of DNA to work with.

The idea of sequencing the whole human genome was proposed by an American researcher, Robert Sinsheimer, in 1985. This led to the establishment of a project under the auspices of the US National Institutes of Health (NIH) in the early 1990s. But one of the NIH researchers, Craig Venter, left to set up his own genome project, using a slightly different technique. This was a commercial venture under the name Celera. The result was a bitter rivalry and a race to map the human genome, which culminated in 2001 in a brief truce when the two groups, each involving thousands of researchers, published their maps. By agreement, Venter's team published in the journal *Science*, and the NIH team, by now known as the International Human Genome Sequencing Consortium, in the journal *Nature*, both in the second week of February 2001.

In truth, neither map was complete. Both teams recognized that their sequences were first drafts; each of them had sequenced something over 90 per cent of the human genome, with gaps about 100,000 letters long. The rivalry to complete the task soon became as bitter as ever. Celera now focused on developing drugs based on the genome information, while the Consortium pushed on to plug the gaps. As a result, there were several subsequent claims that the mapping was 'complete', and the NIH in particular prefers 2003 as the landmark date. But as the American commentator Laura Helmuth, who has followed the saga throughout, has pointed out, 'Celebrating 2003 rather than 2001 as the most important date in the sequencing of the human genome is like celebrating the anniversary of the final Apollo mission rather than the first one to land on the moon.'[50]

The number 15 can be obtained by multiplying the number 3 by the number 5; the factors of 15 are 3 and 5. This may not seem a very profound piece of knowledge. But during the early years of the twenty-first century the development of a new kind of computer that could work this out for itself (by 'factorisation' of the number 15) marked a giant leap towards practical quantum computing. A genuine quantum computer would be as far in advance of a conventional computer as a conventional computer is in advance of an abacus. But such machines may be no more than twenty years away.

A conventional computer operates by manipulating numbers in binary code, represented as strings of 0s and 1s, which can be thought of as switches that are either on or off. Each switch corresponds to a 'bit' of information, and eight-bit 'words' are known as bytes. One measure of the power of a computer is the number of bits (or number of bytes) that can be stored and manipulated during calculations. But when we are dealing with quantum entities, such as individual atoms or electrons, the rules are different. A quantum 'switch' can be both on and off at the same time, in a so-called superposition of states.

An electron provides an example of such a phenomenon. Electrons have a property called spin, which can be thought of as like an arrow pointing up or down. 'Up' might correspond to 0 in binary language, and 'down' would then correspond to 1. But in the right circumstances an electron can exist in both states simultaneously (as can some atoms).

The result is known as a quantum bit, or qubit (pronounced like the ancient unit 'cubit'). And the power of a string of qubits, in computational terms, is the same as the power of a conventional computer with a number of bits equal to 2 raised to the power of the number of qubits. So a computer built from 4 qubits has the power of a conventional computer with $2 \times 2 \times 2 \times 2 = 16$ bits, and so on. Such exponentials grow very quickly; a quantum computer containing just 10 qubits would have the power of a conventional computer with 1,024 bits (this is known as a kilobit, because it is very nearly 1000 bits).

Unfortunately, it is very difficult to maintain strings of qubits, manipulate them, and read out data from them. But a start has been made, and the factorisation problem was solved as a result.

Factorisation was chosen as one of the first problems to be tackled because it is of vital importance in manipulating the codes used in security by banks, big business, military uses, and to keep information safe on the internet. Codes used in these areas are based on the properties of very large numbers, hundreds of digits long, which are made by multiplying together two large prime numbers (that is, numbers which cannot themselves be factorized, like

7 or 317, but much bigger). The very large number is used to scramble the message being coded, which can then be unscrambled only by someone who knows these factors. The security comes because it is very difficult to find the factors of a very large number. But what might take years on a conventional computer could be done in minutes on a full-sized quantum computer. The technique for doing this was developed by Peter Shor, of Bell Labs, and is known as Shor's algorithm.

In 2001, a team of IBM researchers managed to manipulate a system containing molecules each made up of five fluorine atoms and two carbon atoms to act like a 7-qubit computer. This is equivalent to a conventional computer with 2^7 bits (128 bits). They used this to test Shor's algorithm, by using the 'computer' to work out the factors of 15. You will not be surprised to learn that they got the answer 3 and 5.

Unfortunately, the drawback of this particular technique is that it cannot be scaled up to large numbers of qubits; but it provided clear proof that Shor's algorithm works and that a scaled-up quantum computer would be a powerful machine that would not only require an overhaul of security systems but could tackle many other problems.

A key breakthrough came in 2012, when a team at the Santa Barbara campus of the University of California (UCSB) ran a version of Shor's algorithm on a solid state processor containing four superconducting phase qubits, the quantum equivalents of transistors. Like their predecessors, they found the factors of the number 15, the smallest number that Shor's technique can be applied to. But unlike the IBM work, this really was quantum computing on a chip, and has the potential to be scaled up to something much larger. It won't be easy, but as Andrew Cleland, one of the UCSB researchers said at the time, 'the path forward is clear.'

Peter Shor (b. 1959).

MAKING MATTER MASSIVE

T he idea behind what became known as the Higgs particle emerged in the early 1960s, from studies of the way the forces of nature behave at a subatomic level. This was very much an idea whose time had come, and in 1964 several teams of researchers were investigating the possibilities. Peter Higgs worked alone at the University of Edinburgh; François Englert and Robert Brout worked together in Belgium; Tom Kibble, Gerald Guralnik, and Carl Hagen were working at Imperial College in London. Each team independently published papers pointing out more or less the same thing. Unknown to any of them, two young Russian theorists, Sacha Migdal and Sacha Polyakov, had also come up with the idea, but had been dissuaded from publishing by sceptical senior scientists who dismissed the suggestion.

The big idea concerns the behaviour of the particles that carry what is known as the weak force of nature. These are analogous to photons, the particles of light, but unlike photons they have mass. The weak force is involved in things like the nuclear interactions that keep the Sun shining, so all this is of far more than academic interest. But why should the particles of the weak force (dubbed W and Z bosons) have mass, while photons do not?

What became known as the BEH mechanism, from the initials of Brout, Englert and Higgs, offered an explanation, first pointed out by Kibble. But along the way it also offered an explanation of where the mass of every particle comes from, including things like electrons, and the quarks that seed the protons and neutrons that make up our everyday material world.

According to this idea, the Universe is filled with a field that interacts more strongly with some families of particle than with others. You can make a rough analogy with a magnetic field. Magnetic objects moving in a magnetic field are affected, but non-magnetic objects sail through the field serenely unaware of its existence. The only objects that are unaware of what is now called the Higgs field are photons, and they zip through it at the speed of light. Everyone knows that it is harder to get a heavy object moving than to get a light object moving. This is because particles that 'feel' this field a little bit move nearly as fast as photons at the slightest 'push', and particles that 'feel' the field strongly move correspondingly more slowly for a given amount of push. They have more inertia, which corresponds to more mass. But the field itself does not change. It is a bit like trying to wade through water. A streamlined fish can move much faster than a wading person, although they are both moving through the same liquid. This interaction with the Higgs field is what gives particles the property we call mass.

Alone among the six pioneers who published this idea, Peter Higgs realized that the field must have its own boson, and predicted its properties. This boson became known as the Higgs particle. Hardly surprisingly, the predicted Higgs particle would be aware of its own field, so it had to have mass. Such particles would not be around naturally today (although they were in the Big Bang), having given up their energy and decayed, in a process like radioactive decay. But they could be manufactured out of pure energy, in line with Einstein's famous equation, in powerful enough machines. The mass predicted for the particle was so big, however, that there was no chance it could be detected with the particle accelerators available in the 1960s.

It was only in July 2012, 48 years after the mechanism was proposed, that it was announced that experiments at CERN's Large Hadron Collider had observed a new type of particle in the mass region around 126 GeV. The properties of the particle exactly matched the predicted properties of the Higgs boson, in the simplest version of the BEH mechanism. This was

Artist's impression of a proton-proton collision creating bosons.

One end of the CMS (Compact Muon Solenoid) detector at CERN's Large Hadron Collider, during maintenance.

an astounding achievement by what has been described as the most extensive (and expensive!) complex precision instrument in history, involving hundreds of researchers working on a project lasting more than two decades. The importance of the experiment, as much as the importance of the theory, was recognized in 2013, when the Nobel Prize in Physics was awarded to Higgs and Englert (Brout died in 2011, just before his idea was proved correct), but the experiment was also mentioned in the unusually lengthy citation 'for the theoretical discovery of a mechanism that contributes to our understanding of the origin of mass of subatomic particles, and which recently was confirmed through the discovery of the predicted fundamental particle, by the ATLAS and CMS experiments at CERN's Large Hadron Collider'.

N°. 100 THE COMPOSITION OF THE UNIVERSE

When Archimedes took his famous bath (see page 16), the Sun was thought to go round the Earth, and the stars were thought to be points of light attached to a sphere not much further away than the Sun. Now, we know the composition of the Sun and stars, and recent experiments have revealed that these are just minor constituents of a Universe composed of exotic 'dark matter' and 'dark energy'. Along the way, the age of the Universe (the time since the Big Bang) has been pinned down as 13.8 billion years.

Many observations have contributed to these discoveries, not least the studies of distant supernovae described on page 261. But the ultimate experiment (so far) was carried on board the satellite Planck (named after the man who first explained the nature of blackbody radiation), and the results were announced early in 2015. The satellite had been launched by the European Space Agency (ESA) on board an Ariane 5 rocket on 14 May 2009. It spent years observing the sky before the first results were announced in March 2013. Observations still continued until October that year, when the satellite was switched off as it had run low on fuel. But the analysis of the data continued up until the beginning of 2015, when the definitive conclusions were announced.

As its name suggests, the instruments on board Planck were designed to study in detail the cosmic microwave background radiation discovered half a century earlier by Arno Penzias and Robert Wilson (see page 246). But where Penzias and Wilson could detect only a seemingly uniform hiss of radio noise coming from all directions in space, Planck could measure tiny differences in the temperature of the sky from place to place, anisotropies that revealed the state of the Universe at the time the radiation 'decoupled' from matter, a few hundred thousand years

after the Big Bang. This happened when the Universe was cool enough for positive nuclei and negative electrons to join together to make neutral atoms, at which point matter stopped interacting with electromagnetic radiation. Some of the photons detected by Planck came from the regions in which matter was more densely packed and became the seeds from which galaxies and stars developed; others came from regions in the gaps between these seeds of galaxies. Crucially, this corresponds to temperature differences.

The nature of these anisotropies, or 'ripples', in the background radiation, depends on a balancing act in the early Universe. Matter (a combination of dark matter and the stuff we are made of, baryons) tries to pull things together by gravity. But radiation, in the form of energetic photons, tries to smooth things out. The interaction between these effects produces waves, known as acoustic oscillations, with a mixture of wavelengths. When the radiation and matter decouple, the pattern of these

Planck and Herschel launch, artwork. These two missions were launched into space on 14 May 2009 by an Ariane 5 rocket. The spacecraft are mounted on top of the rocket's second (upper) stage. Here, the covering (fairing) has separated, revealing the spacecraft inside. First Herschel (seen here), and then Planck (inside its carrier, black), separated from the upper stage and travelled away from Earth to carry out their missions.

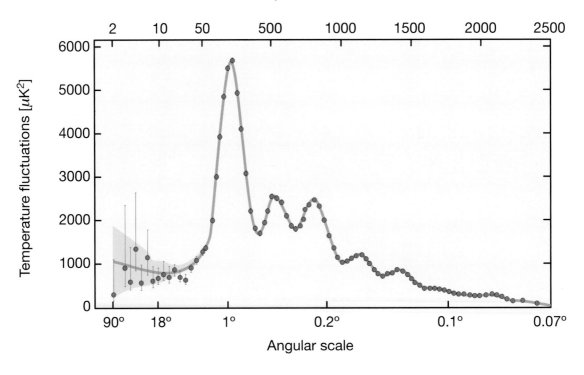

Multipole moment, ℓ

Power spectrum analysis of Planck data (dots) exactly matches the curve predicted by the model of the Universe described in the text – 4.9% baryonic matter, 26.8% dark matter, 68.3% dark energy and an age of 13.8 billion years.

OPPOSITE **Cosmic Microwave Background, as seen by the European Space Agency's Planck satellite (upper right half) and by its predecessor, NASA's Wilkinson Microwave Anisotropy Probe (WMAP) (lower left half). Planck has delivered the most detailed image so far of the Cosmic Microwave Background Radiation.**

waves is frozen in, in the form of the temperature differences from place to place, which Planck measured. The differences are tiny. In some places, the leftover radiation from the Big Bang is a few millionths of a degree warmer, and in others a few millionths of a degree cooler. But Planck was able to measure the differences, and the experimenters were able to unravel the pattern of waves that produced them, in a process similar to the way in which the individual notes that make up a guitar chord can be unravelled by analysing the sound of the chord. The technique is called power spectrum analysis, and although it is a standard process, the complexity of the analysis was made clear by the researcher who described it as being like working out the structure of a grand piano by analysing the noise it made when it was pushed down a flight of stairs.

After all that effort, the result is that we know the composition of the Universe in astonishing detail. Just 4.9 per cent of the Universe is composed of baryonic matter – what we think of as 'ordinary' matter (the things stars, planets and people are made of). Another 26.8 per cent of the Universe is made up of dark matter, which is ordinary in the sense that it seems to be in the form of some kind of particles, but ones which do not interact with baryonic matter at all except by gravity. And that leaves 68.3 per cent of the content of the Universe (more than two-thirds) in the form of the dark energy discovered by the supernova teams. Along the way, Planck pinned down the age of the Universe as 13.798 billion years (13.8 billion years to one decimal place), and measured the speed with which it is expanding, the Hubble constant, as 67.8 kilometres per second per Megaparsec. All of this agrees very well with measurements made by a previous satellite, known as WMAP, but WMAP's results were a little less precise, making the Planck experiment the benchmark for our understanding of the Universe.

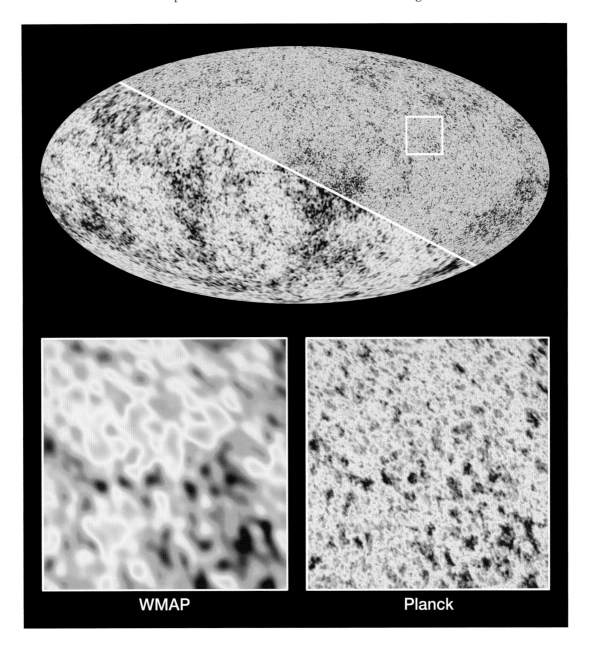

WMAP

Planck

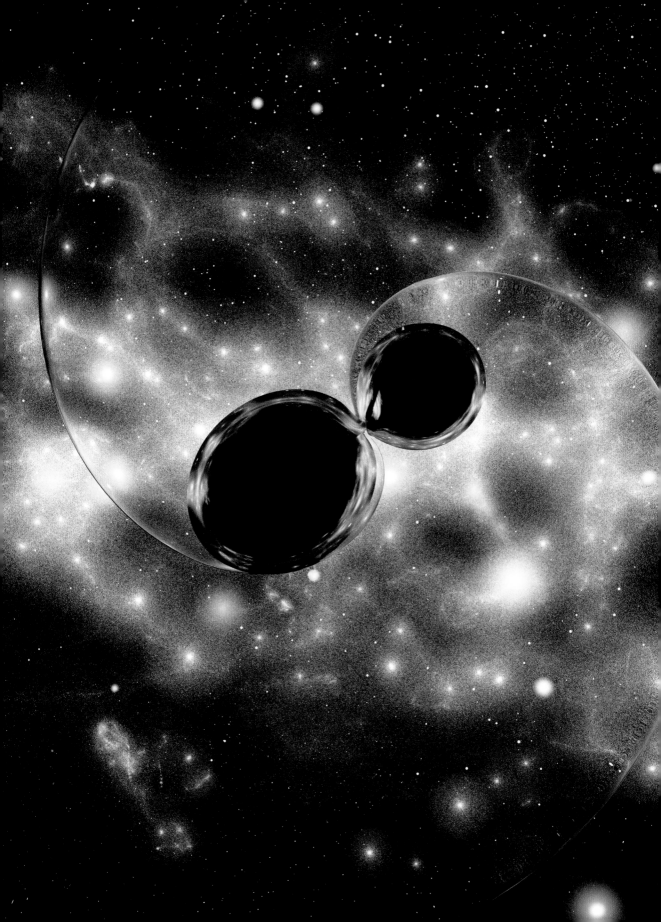

EXPERIMENT 101

I n February 2016, a large team of scientists reported the results of an experiment that so beautifully encapsulates Feynman's description of the scientific method that it deserves to be known as 'Experiment 101', in the same way that basic courses at university are often called, for example, 'Physics 101'. The computation in this case comes from the general theory of relativity, which tells us that massive objects orbiting one another in space should produce ripples in space itself, gravitational waves, that spread out across the universe. We already had evidence that such waves exist, from studies of objects known as binary pulsars (see page 248). But if they could be detected directly in experiments here on Earth that would give us a new way of looking at the Universe.

The way to do this (as Feynman would have said, the way to test the law) involved building a detector with two arms, at right angles to one another, made of evacuated tubes in which laser beams could travel four kilometres down the tube, bounce off a reflecting surface attached to a very heavy mass suspended from a system of pendulums, and then back down the tube to be combined with the other laser beam. The system was set up with exquisite precision, shielded from outside disturbances, with the two laser beams marching in step (actually, exactly out of step!) so that when they combined they exactly cancelled each other out.

The prediction (Feynman's 'guess') was that a gravitational wave passing through this experiment would stretch and squeeze the two right-angle beams in different ways, so that the lasers got out of step and produced a flicker of activity in the detectors where they combined. The calculation, using the general theory of relativity, predicted exactly what kind of pattern would be produced. And on 14 September 2015 exactly that pattern was seen in a pulse lasting just under a tenth of a second, corresponding to a change in the lengths of the laser beams of less than the diameter of an atom.

Even better, the pulse was seen in two identical detectors, one on each side of the North American continent, with a delay of just 6.9 milliseconds between its arrival at the first detector and its arrival at the second detector. This confirmed that it was real, and that it travelled at the speed of light. The exact pattern of ripples in the pulse matches the predictions for the collision and merger of two black holes, each with about 30 times the mass of our Sun; in the process, about three times the mass of our Sun was converted into energy in the form of gravitational waves, in line with Einstein's famous equation $E = mc^2$. This enormous outburst of energy, from a source estimated as being slightly more than a billion light years away, was able to shake detectors on Earth by a tiny amount.

This astonishing experimental result was the culmination of more than two thousand years of experimental science. And it all began with another kind of wave – ripples in the bathtub of a Greek philosopher called Archimedes.

Artistic impression of the merger of two black holes (centre) causing gravitational waves (spiralling lines). In September 2015, gravitational waves were detected directly for the first time. The waves emanated from the collision of two black holes, of 36 and 29 solar masses, some 1.3 billion light years away. The waves were detected by the two LIGO detectors in the USA.

REFERENCES

1. Richard Feynman, *The Key to Science*, Lecture at Cornell University, 1964 (www.youtube.com/watch?v=b240PGCMwV0)

2. William Gilbert, *On the Loadstone and Magnetic Bodies, and On the Great Magnet the Earth*, trans. P. Fleury Mottelay (New York: John Wiley and Sons, 1893)

3. P. A. M. Dirac, *The Principles of Quantum Mechanics*, 1893 (Oxford University Press, Oxford, 1958)

4. Galileo Galilei in *Dialogue Concerning Two New Sciences 1638* (Prometheus Books 1991)

5. Letter from Périer to Pascal, 22 September 1648, in Blaise Pascal, *Oeuvres complètes* (Paris: Éditions du Seuil, Paris, 1964)

6. Royal Society archive: 'Letter of Benjamin Franklin Esq. to Mr. Peter Collinson F. R. S. concerning an Electrical Kite. Read at R.S. 21 Decemb. 1752. Ph. Trans. XLVII. p. 565'

7. Joseph Black and John Robison, *Lectures on the Elements of Chemistry* (Longman and Rees, London, 1803)

8. Quoted in Andrew Carnegie, *James Watt* (Doubleday, Page & Company, New York, 1905)

9. Antoine Lavoisier, *Mémoires of the French Academy* (1786)

10. J. L. E. Dreyer (ed.), *The Scientific Papers of Sir William Herschel*, 2 vols (The Royal Society, London, 1912)

11. Clifford Cunningham, 'William Herschel and the First Two Asteroids' in *The Minor Planet Bulletin* (Dance Hall Observatory, Ontario, 11:3, 1984)

12. J. J. Berzelius, *Essai sur la théorie des proportions chimiques* (Paris, 1819)

13. Lucretius, *The Nature of Things* (trans. A. E. Stallings, Penguin Books, London, 2007)

14. Charles Darwin, *Voyage of the Beagle* (Penguin Books, London, 1989; first published 1839)

15. *Ibid.*

16. Louis Agassiz, '*Discours prononce a l'ouverture des seances Société Helvétique des Sciences Naturelles*', address delivered at the opening of the Helvetic Natural History Society, at Neuchâtel, 24 July 1837 (*The Edinburgh New Philosophical Journal*

v.24, Oct. 1837–April 1838, pp.364–383)

17. Louis Agassiz, *Études sur les Glaciers* (Jent et Gassmann, Neuchâtel, 1(3) 122, 1840)

18. John Tyndall, *Light and electricity*, notes of two courses of lectures before the Royal Institution of Great Britain (D. Appleton and Co., New York, 1883)

19. Thomas Lefroy, *Memoir of Chief Justice Lefroy* (Hodges, Foster & Co., Dublin, 1871)

20. Crawford Long, An account of the first use of Sulphuric Ether by Inhalation as an Anaesthetic in Surgical Operations (*Southern Medical and Surgical Journal*, vol. 5, 705–713, 1849)

21. Letter to the editor of the *Medical Times and Gazette*, September 1854

22. *Ibid.*

23. Louis Pasteur, *Methode pour prevenir la rage apres morsure* (C. R. Acad. Sci. 101, 765–774, 1885) quoted at: www.historylearningsite.co.uk/a-history-of-medicine/louis-pasteur/

24. Reprinted in R. S. Shankland, 'Michelson–Morley experiment' (*American Journal of Physics*, 31(1), 1964)

25. Quoted at: http://web.archive.org/web/20090925102542/http://chem.ch.huji.ac.il/history/hertz.htm

26. Lord Rayleigh, 'The Density of Gases in the Air and the Discovery of Argon', *Nobel Lecture*, 12 December 1904

27. Sir William Ramsay, 'The Rare Gases of the Atmosphere', *Nobel Lecture*, 12 December 1904

28. J. J. Thomson, 'On the Masses of the Ions in Gases at Low Pressures' (*Philosophical Magazine*, 5:48, No.295, pp.547–567 (page 565), December 1899)

29. Antoine H. Becquerel, 'On radioactivity, a new property of matter', *Nobel Lecture*, 11 December 1903

30. Philipp E. A. Lenard, 'On Cathode Rays', *Nobel Lecture*, 28 May 1906

31. Robert A. Millikan, 'The electron and the light-quant from the experimental point of view', *Nobel Lecture*, 23 May 1924

32. Ivan Pavlov, 'Physiology of Digestion', *Nobel Lecture*, 12 December 1904

33. *Ibid.*

34. William Lawrence Bragg, 'The diffraction of X-rays by crystals', *Nobel Lecture*, 6 September 1922

35. Clinton J. Davisson, 'The discovery of electron waves', *Nobel Lecture*, 13 December 1937

36. Alexander Fleming, 'Penicillin', *Nobel Lecture*, 11 December 1945

37. *Ibid.*

38. Albert Szent-Györgye, 'Oxidation, energy transfer, and vitamins', *Nobel Lecture*, 11 December 1937

39. Dorothy Crowfoot Hodgkin, 'X-Ray Photographs of Crystalline Pepsin', in *Nature*, 133, 795, 1934

40. Frédéric Joliot and Irène Joliot-Curie, 'Artificial Production of Radioactive Elements', *Nobel Lecture*, 12 December 1935

41. Frédéric Joliot, 'Chemical evidence of the transmutation of elements', *Nobel Lecture*, 12 December 1935

42. Otto Hahn, 'From the natural transmutations of uranium to its artificial fission', *Nobel Lecture*, 13 December 1946

43. Enrico Fermi, 'Fermi's Own Story', in *The First Reactor* (United States Department of Energy, DOE/NE-0046, pp.25–26, 1982)

44. Linus Pauling, *The Nature of the Chemical Bond* (Cornell UP, 1939)

45. Milly Dawson, 'Martha Chase dies', *The Scientist*, 20 August 2003

46. From the University of Cambridge Darwin Correspondence Project: http://www.darwinproject.ac.uk/letter/DCP-LETT-7471.xml

47. Dorothy Crowfoot Hodgkin, 'The X-ray analysis of complicated molecules', *Nobel Lecture*, 11 December 1964

48. Robert W. Wilson, 'The Cosmic Microwave Background Radiation', *Nobel Lecture*, 8 December 1978.

49. J. D. Hays, John Imbrie, N. J. Shackleton, Variations in the Earth's Orbit: Pacemaker of the Ice Ages' (*Science*, 194: 4270, pp. 1121–1132, 10 December 1976).

50. Laura Helmuth, 'Watch Francis Collins Lunge for the Nobel Prize', *Slate*, 4 November 2013

INDEX

Ceres 74, 75–6
Chadwick, James 197–200
chain reaction 207–8
Charles, Jacques 76–7, 77, 85
Charles's law 77
Charlière balloons 76–8
Chase, Martha 220–1, 221
chemical reactions 46–7, 59–60
    Dalton's atomic theory 85
    electrolysis 86–7
chemical symbols 59, 85, 88, 89
Chicago Pile 1 (CP1) 207–8
chlorine 238
chloroform 122
cholera 120–2
chromosomes 173, 176, 267
    maize plants 215–17
circumference of Earth 18–19
Cleland, Andrew 269
clocks on a plane experiment 246–7
Cockcroft, John 192
code-breakers 209
coevolution 130–1
coherent radiation 231
Collinson, Peter 44
Colossus computers 209–13
combustion 58–9
comets 57, 229
compounds 85, 88–90
    aromatic 132–4
    synthesis 97–9
computers 209–13, 241
    quantum computing 268–9
conditioned response 162
conservation laws 60, 66, 104, 105–6
controlled experiments 43, 71, 115–16
Cook, James 43
Copenhagen Interpretation 202–3
core of the Earth 25, 163–5
Corey, Robert 218
cosmic distance scale 169–70
cosmic microwave background
        radiation 244–6, 274–6
cosmological constant 264
Couper, Archibald Scott 132–3
Cowan, Clyde 236–8
cowpox 69–70
Crab Nebula 114
Creighton, Harriet 215
Crick, Francis 225–6
Croll, James 253

Crookes tube experiment 137–8, 150–1
Crookes, William 137
crystals 193, 195
    diffraction of electrons 182, 183
    diffraction of X-rays 177–8
    DNA 224, 225
    phosphorescence 154–5
    piezoelectric effect 138–40
    structure 37, 178–9
    see also X-ray crystallography
Curie, Jacques 139
Curie, Marie 156–7
Curie, Pierre 139, 156–7
cytoplasm 170, 171

D

d'Alibard, Thomas-François 44, 45
Dalton, John 82–5
Dalton's atomic theory 82–5
Dalton's law 83
dark energy 264, 276–7
dark matter 276
Darwin, Charles 102–4
    natural selection 13–14, 103,
        129–31
Davis, Raymond 236, 238
Davisson, Clinton 182
Davy, Humphry 51, 85–7
de Broglie, Louis 181, 182–3
De Magnete 10, 24–6
De Motu 28–30
deoxyribonucleic acid see DNA
Descartes, René 30, 31, 32
Diesel, Rudolf 92
diffraction 82
    electrons 182, 183
    X-rays 177–8
    see also double-slit experiment;
        X-ray crystallography
digestion 159, 162
Dirac, Paul 14–15
displacement of a fluid 16–17
dissection 21–4, 61, 130
DNA 172, 222
    discovery 170–2
    role of 213–15, 219–21
    sequencing 265–7
    structure 224–6
dogs, Pavlov's 159–62
Doppler, Christian 106
Doppler effect 106–7

double helix, discovery of 224–6
double-slit experiment 79–82
    individual electrons 259–61
dust motes 92

E

Earth
    age of 102
    axial tilt 19, 253
    breathing 241–3
    core 25, 163–5
    density 65, 163, 165
    early 227–8, 229
    geological observations 102–4
    inner structure 25, 103–4, 163–5
    magnetic field 24–6, 78, 165, 232–3
    measuring size of 18–19
    orbital geometry 253–4, 255
    plate tectonics 232–3
    temperature 112
    weighing the 63–5
earthquakes 102, 103, 163–5
eclipses 40–1, 179–81
Eddington, Arthur 180–1
efficiency 49–50, 90–2
Einstein, Albert 207, 230, 259
    on Brownian motion 94
    EPR Paradox 258
    equation 206, 279
    photoelectric effect 158–9
    see also general theory of relativity;
        special theory of relativity
electric arc lamp 86, 87
electric motor 97, 99
electricity 44–6, 94–7
    in animals 61–2, 63
    and chemistry 85–7
    generation 63, 99–101, 208
    in a vacuum 136–8
electrodynamics 94–7
electrolysis experiments 85–7
electromagnetic radiation 143–4
electromagnetism 99–101, 143
    discovery of 94–7
    Maxwell's equations 123, 126, 140
electrons 167–8, 230, 268
    discovery 152–4
    double-slit experiment 259–61
    photoelectric effect 154, 157–9
    wave-particle duality 181–3, 259–61
    see also cathode rays

ACKNOWLEDGEMENTS

The authors thank the Alfred C. Munger Foundation for financial support, and the University of Sussex for providing a base from which to work.

Picture credits

The publishers would like to thank the following for their kind permission to reproduce images:

5 NASA/Science Photo Library; 8–9 Caltech/MIT/Ligo Labs/Science Photo Library; 10 National Library of Medicine/Science Photo Library; 11 Physics Today Collection/American Institute of Physics/Science Photo Library; 12 Science Source/Science Photo Library; 13 Paul D. Stewart/Science Photo Library; 16 Science Photo Library; 18 Sheila Terry/Science Photo Library; 19 Collection Abecasis/Science Photo Library; 20 Science Museum/Science & Society Picture Library; 21 Science Source/Science Photo Library; 22 British Library/Science Photo Library; 23 British Library/Science Photo Library; 25 Science Photo Library (×2); 27 New York Public Library/Science Photo Library; 29 Dr Jeremy Burgess/Science Photo Library; 30 Science Photo Library; 31 Science Source/Science Photo Library; 33 Royal Astronomical Society/Science Photo Library; 35 British Library/Science Photo Library; 36 Natural History Museum, London/Science Photo Library; 37 David Parker/Science Photo Library; 38 Universal History Archive/UIG/Science Photo Library; 41 New York Public Library/Science Photo Library; 42 St. Mary's Hospital Medical School/Science Photo Library; 43 Science Photo Library; 44 Print Collection, Miriam and Ira D. Wallach Division of Art, Prints and Photographs/New York Public Library/Science Photo Library; 45 Sheila Terry/Science Photo Library; 46 Middle Temple Library/Science Photo Library; 47 Sheila Terry/Science Photo Library; 49 Claus Lunau/Science Photo Library; 50 Science Photo Library; 52–53 Science Photo Library; 54 Biophoto Associates/Science Photo Library; 56 New York Public Library Picture Collection/Science Photo Library; 57 Royal Astronomical Society/Science Photo Library; 58 Science Photo Library; 60 Gregory Tobias/Chemical Heritage Foundation/Science Photo Library; 61 Science Photo Library; 62 Science Source/Science Photo Library; 64 Dorling Kindersley/UIG/Science Photo Library; 66 Science Photo Library; 67 Sheila Terry/Science Photo Library; 68 Jean-Loup Charmet/Science Photo Library; 69 CDC/Science Photo Library; 70 Doublevision/Science Photo Library; 72 Universal History Archive/UIG/Science Photo Library; 74 NASA/JPL-Caltech/UCLA/MPS/DLR/IDA/Science Photo Library; 75 Science Photo Library; 77 Sheila Terry/Science Photo Library; 79 Emilio Segre Visual Archives/American Institute of Physics/Science Photo Library; 80–81 Erich Schrempp/Science Photo Library; 82 GIPHOTOSTOCK/Science Photo Library; 83 David Taylor/Science Photo Library; 84 Science Photo Library; 86 Science Photo Library; 87 Sheila Terry/Science Photo Library; 89 Sheila Terry/Science Photo Library; 90 Ashley Cooper/Science Photo Library; 93 Patrick Dumas/Look at Sciences/Science Photo Library; 94 American Philosophical Society/Science Photo Library; 95 GIPHOTOSTOCK/Science Photo Library; 96 Sheila Terry/Science Photo Library; 98 Molekuul/Science Photo Library; 99 Sheila Terry/Science Photo Library; 100 Royal Institution of Great Britain/Science Photo Library; 101 Royal Institution of Great Britain/Science Photo Library; 102 Sheila Terry/Science Photo Library; 103 Natural History Museum, London/Science Photo Library; 104 Photo Researchers/Science Photo Library; 105 Science Museum/Science & Society Picture Library; 107 Jose Antonio Peñas/Science Photo Library; 108 British Library/Science Photo Library; 109 Getty Images: Print Collector/Contributor; 110 Royal Institution of Great Britain/Science Photo Library; 111 British Library/Science Photo Library; 113 David Parker/Science Photo Library; 114 Royal Astronomical Society/Science Photo Library; 116 Science Photo Library; 118 Emilio Segre Visual Archives/American Institute of Physics/Science Photo Library; 118 Detlev Van Ravenswaay/Science Photo Library; 119 Science Photo Library; 120 National Library of Medicine/Science Photo Library; 121 Science Photo Library; 123 Science Photo Library; 124–125 National Physical Laboratory © Crown Copyright/Science Photo Library; 127 Science Photo Library; 129 Sinclair Stammers/Science Photo Library; 131 Natural History Museum, London/Science Photo Library; 132 Alfred Pasieka/Science Photo Library; 133 Science Photo Library; 134 Martyn F. Chillmaid/Science Photo Library; 135 James King-Holmes/Science Photo Library; 137 Universal History Archive/UIG/Science